Practical CM

Practical CM

Best Configuration Management Practices

David D Lyon

OXFORD   AUCKLAND   BOSTON   JOHANNESBURG   MELBOURNE   NEW DELHI

Butterworth-Heinemann
Linacre House, Jordan Hill, Oxford OX2 8DP
225 Wildwood Avenue, Woburn, MA 01801-2041
A division of Reed Educational and Professional Publishing Ltd

⊄ A member of the Reed Elsevier plc group

First published by Raven Publishing Company 1996
Second edition 1999
First published in the UK 2000

British Library Cataloguing in Publication Data
A catalogue record for this book is available from the British Library

Library of Congress Cataloguing in Publication Data
A catalogue record for this book is available from the Library of Congress

ISBN 0 7506 4724 8

Composition by Genesis Typesetting, Rochester, Kent
Printed and bound in Great Britain by Biddles Ltd, www.biddles.co.uk

PLANT A TREE

British Trust for Conservation Volunteers

FOR EVERY TITLE THAT WE PUBLISH, BUTTERWORTH-HEINEMANN
WILL PAY FOR BTCV TO PLANT AND CARE FOR A TREE.

Contents ──────────────────

Preface

This book is intended to provide you with 'best CM practices' for today's and tomorrow's business environments. Its second major objective is to show you how to transition from paper configuration and data management (CM/DM) systems to automated, electronic CM/DM systems.

I believe that I have documented in these pages the information necessary for you to establish appropriate CM/DM controls for the effective management of your designs and the efficient control of your hardware and software products. The methodologies, processes and procedures provided herein will also assure that your business remains compliant with current and future customer requirements, while operating in a cost-effective manner for your entire product life cycle. These systems have been implemented in a major operating unit of a large international corporation and have withstood the test of real-world operation.

The format and presentation of this information is somewhat different from other books dealing with the subject of CM/DM. I hope you will find it enjoyable to read and easy to digest.

The enclosed CD ROM includes a series of templates to fill in and print out. They will form the overall set for your CM/DM system.

David Lyon

Acknowledgements _____

Acknowledgements are in order for the folks whose names I will mention here plus the many others who have helped me throughout my professional career and personal life. I wish I could list you all in these pages but that is not possible.

Special thanks are due my wife, Kathleen, and my daughter, Jennifer, who provided constant encouragement when it was most needed and who accepted my absence from family activities over the years it took to complete this book.

My sincere appreciation goes out to John Williamson and Ron Magee of United Technologies Corporation who showed me what it takes to 'get the job done' in the project engineering and configuration management fields.

I would also like to thank Evan Stutz, my drafting supervisor from my long past engineering apprentice days at the General Electric Company for teaching me the importance of 'paying attention to detail'.

Thank you all.

Abbreviations and acronyms _____

Many of the following acronyms and abbreviations are used throughout this book. They are presented alphabetically in this listing for ease of reference. Detailed definitions are presented in the text where appropriate. *Note*: Several of the acronyms presented are not used in this book. They are provided here because there is a high probability you will run across them in the course of performing your CM activities.

AI	Action Item
AT	Acceptance Test
ASIC	Application Specific Integrated Circuit
BI	Business Item
CAC	Corrective Action Committee
CAD	Computer Aided Design
CALS	Continuous Acquisition and Logistics Support
CCB	Configuration Control Board
CCI	Contractual Configuration Item
CCITT	Consultative Committee of International Telegraph and Telephony
CDRL	Contract Data Requirements List
CD ROM	Compact Disk Read Only Memory
CGM	Computer Graphics Metafile
CI	Configuration Item
CITIS	Contractor Integrated Technical Information Services
CM	Configuration Management
CMP	Configuration Management Plan
CN	Change Notice
CND	Change Notice Disposition
COTS	Commercial Off The Shelf
CPS	Consolidated Purchasing System
CSA	Configuration Status Accounting
CSCI	Computer Software Configuration Item
CVA	Configuration Verification Item
DES	Design
DFT	Drafting

DI	Data Item
DID	Data Item Description
DM	Data Management
DoD	Department of Defense
Draft_Lib	Drafting Library
DWG	Drawing
ECP	Engineering Change Proposal
EIA	Electronics Institute of America
ENG	Engineering
EPL	Engineering Parts List
E-PROM	Erasable Programmable Read Only Memory
FAI	First Article Inspection
FCA	Functional Configuration Audit
FW	Firmware
HW	Hardware
IGES	Initial Graphics Exchange Specification
IS	Interim Standard
ISO	International Standards Organization
LAN	Local Area Network
LCM	Life Cycle Module
MAC	Macintosh Personal Computer
MB	Megabyte
MFG	Manufacturing
MRB	Material Review Board
MRT	Material Review Team
MRP	Material Resource Planning
MIL SPEC	Military Specification
MIL STD	Military Standard
MS	Microsoft Corporation
NFS	Network File System
NOR	Notice of Revision
PBL	Product Baseline
PC	Personal Computer
PCA	Physical Configuration Audit
PDF	Portable Data Format
PDM	Product Data Manager
PL	Parts List
PRB	Problem Resolution Board
PROM	Programmable Read Only Memory
QA	Quality Assurance
RAM	Random Access Memory
RD	Revision Directive
REV	Revision
RFD	Request for Deviation
RFP	Request for Proposal

RFW	Request for Waiver
RI	Responsible Individual
SCI	Software Configuration Item
SCN	Specification Change Notice
STD	Standard
SW	Software
TCI	Technical Configuration Item
TCP/IP	Transport Control Protocol/Internet Protocol
TDP	Technical Data Package
TIFF	Tagged Input File Format
W/D	Waiver/Deviation
WIP	Work in Progress
WL	Wire List
WORKLOC	Work Location
XTERM	External Terminal

Introduction _____

Configuration management (CM) today is a struggle, both for those who are trying to impose some degree of control over the design, production and support phases of programmes and for those who are trying to resist CM in a misguided attempt to save time and money.

Each element of CM, i.e., identification, change control, status accounting, and audits, is inexorably linked to, and interwoven with, engineering design methodologies plus quality assurance inspections and audits and manufacturing production processes, no matter how simple or complex the programme, and is thus integral to the process.

Trade-offs exist every step along the way. These trade-offs involve cost versus control of the design and visibility into how the hardware and software products relate to the design at any given time.

Manual, labour intensive CM activities involving baseline capture and control with change approval and incorporation processes employing multiple forms, databases and meetings induce images of wasteful, expensive pillaging of programme coffers to programme and functional managers.

Thus proposals are often trimmed of CM-related quotes and activities even before the programme or project begins. This usually results in additional costs down the line from excessive changes to the design package, non-conforming hardware, repairs to or reworking of the product hardware and software, and failures experienced in product performance after delivery to the customer.

Take heart, though. Help is on the way!

For each specific CM activity, we will first examine the conventional, classical CM as it is practised today in large businesses and corporations. Next we will implement, in a step-by-step manner, those processes necessary to achieve our ultimate goals. Many companies are in some stage of learning about new, automated CM processes or have begun an earnest foray into the procurement of a Product Data Manager (PDM) system to solve their problems. We will learn how to get to that state in a safe and sane manner. Automated CM is the way to go but much must be said before the plunge is made.

I believe that by addressing the key CM issues presented in this book and by applying the procedures and guidelines defined herein, your business can reap the rewards of an effective control system and sound risk mitigation techniques.

The journey we will take on our road to excellence in CM will follow a route with a few twists and turns in it but the course is set and the direction is true. This easy to follow 'how-to' guide is designed in such a way as to enable you to implement, in a cost-effective manner, practical configuration management solutions for your business for the twenty-first century.

Chapters 1 to 8 will provide you with the 'best CM practices' for today's business environment. These chapters cover the basics of CM, and describe how CM ought to be practised in businesses of all sizes. Current CM methodology is discussed and then the evolution to a practical, effective and ISO-compliant CM methodology is presented as near-term and long-term solutions.

Chapter 9, supplemented by Appendices B, C and D, will provide you with the complete process for the planning, implementation and integration of a PDM system in your business. You will also learn how to integrate the 'best CM practices' presented in Chapters 1 to 8 into your business processes. You will be aided by numerous detailed figures. Chapter 10 will then summarize these 'best CM practices' and the PDM planning, implementation and integration processes in template format so that you can tailor your new CM system to your specific programme requirements and so that you can be assured that you haven't missed anything along the way.

Appendix A will provide instructions for writing an effective CM plan for today's and tomorrow's business environments. Appendices B and D will provide valuable insight into the complexities of PDM system implementation and integration via 'questions and answers' and 'lessons learned'. Appendix C will guide you through the process of evaluating and selecting PDM products and third-party tools. It will also guarantee that you cover all the bases by showing you how to fill in the unique PDM tool and vendor templates provided therein.

I will attempt to be brief but not skip anything important. By the time you finish this book, you will be able to plan a CM programme sized to your business and be ready to construct detailed templates to assure the achievement of your CM goals.

1
Change philosophy

Everything we do involves change. Configuration Management (CM) is the process of managing change. Design, development, integration, test, production, deployment, delivery, maintenance, and support are all manifestations of the change process.

Without change, we are stagnant. Decay and decline sets in. With change, we strive to something better. We make a better world for ourselves, and we feel the satisfaction of emergence into a better state – unless we mismanage the change process, and that which we have set out to accomplish fails. I listened to this view of the change process presented by Mr Tani Haque at the Automated CM Conference in Tyson's Corners, VA on 1 June 1995. I believe that Mr Haque's vision holds true not only for today but for the future as well.

So, CM – the process of managing change – takes on a new light. It becomes integral to achieving our goals, i.e., a process that is not 'the tail of the dog' but one which serves an equal purpose to the 'higher' disciplines of engineering, manufacturing and management.

Of course, the CM process of which I speak here is not the classical 'green eye shades' occupation where rooms of squinty old men (or CM 'persons') record data by hand in CM logbooks or enter data into out-of-date databases when they are not carrying around forms to be signed and bothering engineers and other important persons for clarifications and approvals.

We will start with that conceptualization of CM under which so many of us have laboured for too long. After all, 'you gotta deal with what's real!'.

From this starting point in the bowels of time and industry, we will first get organized and then move on to the interim states through which we must pass in order to get to the ultimate desired state – transparent CM. In this final blissful state one won't see the old CMers slaving at their menial tasks.

A new CM tool, a Product Data Manager (PDM), will capture baselines and control changes, distribute documentation 'in-place' and save money for businesses and stockholders. Also, a well defined, efficient, workflow-driven CM system will replace the old, time worn reactive process of 'fighting fires'.

This transition won't be as easy as falling off a log, though. There are many steps to be taken and many 'hoops' to be jumped through. There will also be a considerable selling job to be done. This transition is going to be a culture change 'in spades' for many.

Let's start by examining the current processes at work in our businesses.

2
Engineering development and configuration management

If it were up to your engineers, they would work on their designs until they were thoroughly tested, updated and proofed out before they presented them to your CM people for baselining and subsequent formal Configuration Control. After that event, each change presented to their peers and to your management clearly points out the inadequacies of their design (from their perspective). On the other hand, your CM personnel want to put that design under some kind of control from the word go. A tug of war usually ensues.

In today's business environment, neither approach is acceptable or feasible. Shortened design cycles and early transition to production schedules require a more flexible but controlled approach.

Wherever engineering departments exist, a consistent engineering design methodology, integrated with a CM mindset, must be employed, based upon

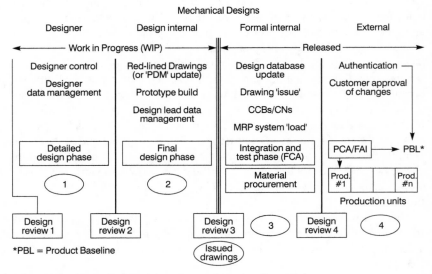

Figure 2.1 Configuration Management levels of control

events which define the transition points from one level of control to another. People must know their roles and responsibilities and understand what is expected of them at specific points in the evolution of the design. The concept of planning engineering or CM milestones by calendar dates just doesn't work because schedules often slip. Event-driven milestones are much more effective. This methodology can be tailored to fit your projects or programmes.

Figure 2.1 presents a methodology for the engineering development process which is adaptable to any programme or product, regardless of the size of your enterprise, provided that you have an engineering organization to start with. If you don't, then use the ideas behind this thought process, anyway. It can't hurt, and you will most likely find that you are headed in the right direction to create order out of chaos.

Design states/levels of control

Four levels of control (as illustrated by four design 'states') are described below:

Designer control

In the first state, 'designer control', the design is initiated and controlled by the designer. Control of the design data is in the hands of the designer. They can use any means at their disposal or within their range of preferences to manage these design data. Your designers' only constraints are those imposed upon them by the system level specification which defines the product requirements. This specification is the output of design review 1. Chapter 3 provides a detailed description of how design reviews should function in your engineering development process and how they should be utilized for CM planning and execution.

Design internal

In the second state, 'design internal', the level of control (still informal) is increased because at that point, an event has occurred which establishes the expectation of a certain amount of 'goodness' in the design, and the design efforts of other people may be impacted by changes to your design as you move into your prototype build phase. In the example presented herein, that event is Design review 2. The designer has enough confidence in their design to sign the drawing and turn it over to your engineering development laboratory to start construction of your prototype hardware.

Proposed changes to the design must be approved by the design lead. It is the lead engineer's responsibility to communicate with other functional areas and personnel proposed changes which may impact their design activities.

As the prototype evolves and design problems and improvements are identified, the initial drawing and/or design tool database is updated, i.e., red-lined drawings and new design files or model versions are created. This iterative process continues until design review 3 is held and all identified problems are resolved.

These two states are generally referred to as 'Work-in-Progress' (WIP).

Formal internal

At the conclusion of design review 3, there is sufficient confidence in the design to justify the procurement of production hardware. This event triggers the transition to the 'formal internal' state.

At this point, the design database is updated, the design data are fed into your business's manufacturing inventory and ordering system, and initial (or all) orders are placed for production hardware.

Drawings which represent the development baseline are issued by your Documentation Control Centre or CM organization (or equivalent) and electronic files generated by design tools which represent the development baseline design are inducted into your PDM system (or not, if you plan to control your design using paper and/or your CAD tools for the near term). This process will be examined and optional solutions proposed in Chapter 4. It is important to note here that, throughout this book, the 'master' design documentation is represented by the design tool database files and not by mylar or paper drawings, as in the past, except in the case of businesses that, for the time being, rely upon paper drawings and specifications to conduct their operations.

Future proposed changes to the design (including changes to the system level specifications) are documented on your internal 'change notices' and are approved or disapproved by your internal Configuration Control Board (CCB). Unless otherwise negotiated, your customer should not participate in the change approval process until the product baseline has been established, as defined below.

If your programme involves military hardware or critical commercial hardware, your customer may request (or demand) to review and initial the drawings before they are issued and inducted into the internal configuration control system. *Note*: This 'initialling' of the documentation is different from the 'authentication' process for military designs. That event does not occur until the product baseline has been established by the satisfactory completion of the Functional Configuration Audit (FCA) and the Physical Configuration Audit (PCA).

The FCA is conducted on the engineering prototype during the 'formal internal' state. The purpose of the FCA is to assure that tests have been conducted to verify that each requirement specified in the system level specification has been met by the design. If tests cannot be performed to verify a particular requirement, then a 'theoretical error analysis' must be performed to verify satisfactory compliance to the requirement. These tests are generally referred to as design evaluation and qualification tests. *Note*: The advent of the terminology FCA and PCA is relatively recent. It is mainly driven by military standards and used primarily on defence programmes. The purpose of these activities is universal, though, and should be an integral part of every development programme.

External

The transition to the 'external' state occurs at the successful completion of the FCA. This event initiates the conduct of the PCA. During the PCA, the engineering drawings are proofed by comparison to the first production unit. This unit should be built to manufacturing planning that was created from the engineering drawing package. Measurements are verified and all instructions, processes and technical data specified on the engineering drawings are verified against the hardware. Historically, many businesses referred to this activity as the 'first article inspection (FAI)'.

The second major activity of the PCA is the capture of drawing revision and serial number data and comparison to 'as-defined' revision levels plus a verification that serial numbers have not been previously used.

The final activity of the PCA is the performance of an Acceptance Test (AT) to assure that the unit examined is physically and functionally the same as the unit that passed the FCA.

Upon resolution of any problems that were observed during the conduct of the FCA and PCA, a FCA/PCA certification form is signed by the customer, and the 'product baseline' is established.

At this point, authentication of the drawings is performed by your customer (if a military contract is involved), and your customer becomes an active participant in the approval of proposed changes.

In commercial businesses, the production run is continued with further confidence in the accuracy of the engineering drawing package and the hardware produced.

Changes continue to be processed by the CCB throughout the life cycle of your product.

3
Tailoring and planning

Tailoring

The general rule in CM operations up to just a few years ago was that the CM organization usually did not get a chance to identify and quote their activities when a business responded to a potential customer's Request for Proposal (RFP). Systems engineering, design engineering, drafting, manufacturing, quality assurance, product support, project and program management got to quote but poor old CM was often left out in the cold, and if, by chance, they did get to quote, they were usually the first to have their budget cut or their efforts (and funding) reduced when the chips were down. When CM activities inevitably became necessary during the conduct of a contract (often because of product failure and/or loss of knowledge of the design or product hardware or software), minimal change control and status accounting activities were usually instituted in a manual, labour intensive manner, and everyone blamed the whole mess on CM – 'how could you have let this happen?'

Fortunately, times have changed, and most businesses that survived the sweep of the 'merge mania' scythe of the late 1980s and 1990s have recognized the value of CM. The importance of capturing design baselines and controlling changes plus having visibility into the status of prototype and production hardware and the performance of formal audits to assure design compliance to system level requirements and to verify that a solid set of build drawings and manufacturing planning exists, has driven customers, both military and commercial, to demand that major aspects of CM be included in proposals for their products.

This new corporate attitude and marketplace mindset has increased the need for a consistent set of CM activities to be identified and quoted at the outset of a contract or programme, regardless of the size of the programme.

These CM activities may be articulated as a compilation of 'best practices' which should be tailored to meet the unique requirements of individual contracts and products. Chapters 4 to 10 present best practices for the CM activities of today plus those of the future and describe the transition process necessary to take us from today's 'paper intensive' CM processes to tomorrow's 'paperless' automated CM. These chapters also describe best practices (electronic or paper) that will reduce risks and provide the appropriate level of control necessary to deliver the product which you promised in your contract or advertising.

Military contracts have, in the past, relied heavily upon and have referenced Military Standards (MIL STD) and Military Specifications (MIL SPEC) to define the scope and level of CM disciplines to be enforced by DoD contractors and suppliers. Procurement reform has initiated the transition of the specification of CM activities from MIL STD and MIL SPEC to commercial guidelines such as the International Standards Organization's ISO 10007 guideline document for CM. The industry and government are currently struggling to maintain control of procurements while this transition is being accomplished. Both groups are also communicating and cooperating in order to arrive at a meaningful and cost-sensitive set of CM disciplines to impose upon themselves.

A meeting in Tyson's Corners, VA on 1–2 June 1995, composed of roughly one hundred government and industry representatives identified the need for the CM community to transition from paper systems to electronic systems such as that called for in the Contractor Integrated Technical Information Services (CITIS) specification (MIL STD 974). We will get into this subject in greater detail in later chapters.

My point here is that customers have needs that they try to articulate in RFPs and document in contracts. Contractors and businesses wish to sell products that meet those needs, perform well and are reliable so that customers come back for more.

The CM templates provided in Chapter 10 provide the basis for successful tailoring of CM 'best practices' to fit the customer's 'real' needs, while satisfying the intent of today's and tomorrow's military and commercial standards. If everyone does their job right, tomorrow's military and commercial standards will make this tailoring effort easier to handle. In any case, this first step, i.e., careful tailoring of requirements, will go a long way to identify those CM activities which give the most 'bang for the buck' and which will provide the basis for proper documentation of customer approval for these CM disciplines.

Planning

The next step in establishing the groundwork for a successful CM programme is the planning activity. I am including 'planning' in the same chapter as 'tailoring' because they are so closely linked. They are built on the same foundation, like the living room and kitchen are built on the foundations of the same house.

If we think of the living room as the location where we invite the customer in to chat about how we are going to apply CM disciplines and processes to meet their requirements, then the kitchen is the place where we cook the stew, i.e., put those processes to work after we assemble our team of cooks.

As in the case of a great meal, the courses require 'tasting' along the way to assure that we haven't strayed from our desired goal. In industry, we call these 'tastings' design reviews.

At each design review, we expect a certain amount of 'goodness' in the design. For the purposes of this book, four design reviews will be examined and integrated into the context of the development activity. The following is a brief description of the expectations for each of these design reviews:

- Design review 1 – The system specification, which defines the system level requirements for the product, has been completed and signed off by your internal functions and your customer. *Your functional baseline (system level) and allocated baselines (subsystem level) are established at this point.*
- Design review 2 – Your designers have developed the detailed design to the point where they are confident that the system level requirements have been met, and they are willing to sign the drawing which represents the product or the part or subsystem in question. At this point, fabrication of the prototype is initiated.
- Design review 3 – The prototype build has been completed, and the design database has been updated to reflect changes made to the design and/or red-lined changes have been made to the drawings. *Your development baseline is established at this point. Note*: It is important to note here that the 'master' of the design is no longer the physical drawing, as in the past, but the 'ones and zeros' in the design database, i.e., the electronic design files, themselves.
- Design review 4 – The Functional Configuration Audit has been successfully completed. This process, defined in detail in Chapter 4, provides documented proof that the requirements specified in the system specification have been met, i.e., the design works.

Various names have traditionally been associated with these design reviews (1 – concept, 2 – detailed or implementation, 3 – critical or pre-release, 4 – final or customer). However, it is more important to identify the specific outputs of the design review as the relevant indication of the 'goodness' of the design at that point in time, rather than the name associated with the design review.

From the CM perspective, it is important to identify CM-related tasks and activities associated with each design review. Figure 3.1 provides a Design Review Checklist for CM. Each activity or procedure indicated must be signed up for by the participating team members prior to the completion of design review 1. In this way, everyone involved knows in advance what is expected of them at each succeeding design review and has 'ownership' of the process in their respective areas, i.e., roles and responsibilities have been assigned and agreed upon.

Project _____

Design review 1:

_____ Determine whether this design review establishes a new baseline or modifies an existing baseline
_____ Determine if new special tooling or test equipment will be built
_____ Ensure that all Configuration Items have been identified
_____ Ensure items to be serialized have been identified
_____ Determine the approximate number of new designs required or existing designs to be modified for this product
_____ Complete CM Planning Schedule – Figure 3.2
_____ Complete CM Planning Checklist – Figure 3.3

Design review 2:

_____ Verify that the CMP is being followed and that CM activities are on schedule
_____ Verify that new special tools and test equipment have been documented
_____ Verify that a list of all new drawing numbers and titles has been provided to CM
_____ Verify that the current configuration identification has been placed under informal internal control

Design review 3:

_____ Verify that the current configuration identification has been placed under formal internal control
_____ Verify that a Configuration Control Board (CCB) has been established for this project
_____ Verify that all digital design data have been updated and that all digital design files and models have been captured and inducted into your Product Data Management (PDM) system vaults
_____ Verify that your business's MRP system has been loaded with accurate configuration data for production orders

Design review 4:

_____ Verify that the FCA has been completed and that all corrective actions have been closed out
_____ Verify that your PDM database is up to date and that all digital design files have been updated

Configuration Management _____ Date _____

Figure 3.1 Design Review Checklist

The CM Planning Schedule is also completed at design review 1. This is the vehicle for recording the Design Review Schedule and, as a result, the schedule for the transition from one level of control to the next. Formal audits are also scheduled and the capture of baselines planned. This planning schedule is shown in Figure 3.2.

Finally, a listing of all inputs to the CM process is presented in Figure 3.3. This list is reviewed with the cognizant design 'lead' as part of the checklist review at the time of design review 1. The required inputs to the CM function

	(1)	Jan	Feb	Mar	Apr	May	Jun	Jul	Aug	Sep	Oct	Nov	Dec
1 Configuration identification													
(a) Baseline events													
* Design reviews													
* Audits													
(b) Capture design													
* Drawings/PL/WL (2)													
* Electronic													
(c) Serialization plan													
2 Configuration control													
(a) Developer													
(b) Design internal													
(c) Formal internal (3)													
(d) External (4)													
3 Status accounting													
(a) MRP loaded													
(b) CM CSA database													
(c) CCB and MRB reports													
4 Formal audits													
(a) FCA													
(b) PCA													
(c) Configuration verification													

Project lead: _____

NOTES: (1) Status column (S): C = Complete; I = In process; L = Late. (2) DFT/Eng/Producibility sign-off. (3) Design issued. (4) Authentication

Figure 3.2 CM Planning Schedule

CM Task	Input (required)	CM process
Configuration identification	Contract requirements	Tailor and negotiate CM activities Generate CM plan
	Drafting planning sheet	Identify baseines plus # of drawings
	Family tree	Show configuration hierarchy for baselines and as-built reports
	Design review schedule	Define transition points for CM control
	Design review checklist	Verify CM activities completed
	Design review report	Capture design baseline
Configuration control	Drawings/design files and pathnames	Put drawings under control Put electronic files and models under control
	Design review schedule	Define levels of control between transition points: *Developer *Design internal *Formal internal *External (customer)
	Design review reports	Initiate CCB activities
	Change paper	Process changes to issued drawings and specifications (TDP)
	Non-conforming material paper (waivers and deviations)	Conduct MRBs (internal and external)
Configuration Status Accounting (CSA)	CCB minutes MRB minutes	Establish CM databases: *MRB (PDM) *CM change paper trail (PDM) *CSA (PDM)
Configuration Audits	Access to product data plus designers, users, builders, maintainers	Support FCA Conduct PCA Generate PCA report Conduct verification audits and generate reports

Figure 3.3 CM Planning Checklist

or organization are identified and related to the CM process which they enable or support.

The thorough resolution and completion of these checklists and schedules provide the basis for the CM programme to be established for your programme or product. A good job here will produce a well designed and reliably built product. A sloppy or incomplete job will spawn a free-wheeling design activity, with a build process that is out of control, i.e., a disaster.

CM Plan

The results of your planning and negotiations should be documented in your CM plan (CMP). Your CMP should show the composition, roles and responsibilities of your Configuration Control Board and demonstrate the manner in which you will implement your CM 'best practices' to fulfil your contract requirements for the major CM areas of:

- Configuration Identification;
- Configuration Control;
- Configuration Status Accounting;
- Configuration Audits.

Chapters 4 to 9 will provide you with procedures and guidelines for establishing the 'best practices' for these CM activities. Chapter 10 provides templates which make it possible for you to develop your own CM programme with the assurance that you have covered all the bases and achieved that indispensable 'buy-in' from those individuals and functions which must support your CM activities in order for your projects and programmes to be successful.

Appendix A will show you how to write a CMP. The important thing to remember is that you must use the methodology discussed earlier in this chapter to assure that your key personnel and functions are aware of their roles and responsibilities early on in the programme and that you have their buy-in of the process. Your CMP will document these agreements and guarantee compliance to your contractual requirements in addition to setting the scene for the effective capture and control of your design baselines. It will guide your organization through the design and development, transition to production, production, delivery, support, and maintenance of your product and its design.

You need only to follow the templates provided in Chapter 10 and the instructions in Appendix A to assure that you have covered all the bases and that you have translated your customer's requirements into a documented set of activities and guidelines plus provided a definition of roles and responsibilities that lines up with the best CM practices as recommended in Chapters 4 to 9. This approach should get you started off on the right foot, regardless of the size of your business.

4

CM basics (today, tomorrow, future)

We will start our journey towards excellence with the CM of today. As I stated in Chapter 1, 'you gotta deal with what's real'. We will identify and explain the purpose of the various elements of CM and examine how companies try to deal with them today.

Next, we will introduce the concept of a Product Data Manager (PDM) and see how this tool, in conjunction with solid CM and engineering development methodologies, will enable us to achieve our goal of automated CM. Chapter 9 will provide the planning and implementation information necessary for the integration of PDM system functionality within our businesses.

Finally, we will examine a safe and sane approach to automated CM via an incremental transition from our current 'paper' CM systems to a 'paperless' CM system as we start off on our journey towards excellence in CM. We will incorporate the technologies and processes with which we will reap the rewards of controlled designs and products while meeting tight schedules at the minimum expense to the programme.

Unfortunately, 'what's real' today is often not the manifestation of what should be real in order to qualify for the 'best practice' award. However, the CM practised at many companies generally follows the four disciplines defined in MIL STD 973 and EIA 649, i.e., Configuration Identification, Configuration Control, Configuration Status Accounting, and Configuration Audits – all of this on paper, of course.

It is appropriate to mention at this point that MIL STD 973 is a much improved version of several older Military Standards. EIA 649 is an improved commercial version of MIL STD 973 and the result of the 'procurement reform' initiative of US Defense Secretary William Perry. The structure of previous MIL STDs contributed greatly to the misunderstanding of what CM is really all about, thereby facilitating the CM 'mess' many companies sloughed into.

MIL STD 973 and EIA 649 both consolidated and clarified the information contained in MIL STD 480 and MIL STD 483 in the area of Configuration Identification and change control. They added sections on software CM and tailoring of CM requirements. EIA 649 also added 'reasons' for various CM requirements. EIA 649 has been issued as the government and industry guidelines for CM.

MIL STD 2549, DoD Interface Standard, Configuration Management Data Interface is the government's product data interface definition document. It identifies the data elements that comprise the Configuration Status Accounting (CSA) product data, provides the corresponding data element definitions, and describes the format and relational structure of these data elements. Data delivered during future contracts between the DoD and industry must conform to this new interface standard.

MIL STD 2549 was fully approved on 30 June 1997 (not as an interim standard, as originally planned) and will be included in RFPs as a required MIL STD for delivery of data on government products.

A test plan has been developed by the MIL STD 2549 Program Manager for a cooperative test program designed to work out any 'bugs' in this document. This test program is being supported by the government, military services and selected industry participants. Training courses on MIL STD 2549 are available from the government.

You can learn more about MIL STD 2549, including available training courses, at the DoD Acquisition and Technology Web Site on the internet at URL: http://www.acq.osd.mil. You can also download a copy of this document plus check on the current status of (and/or download) several other documents relating to the subjects of CM, DM and CALS:

- MIL-HDBK 61: Configuration Management;
- MIL-STD-973: Military Configuration Management;
- MIL-STD-100: DoD Standard Practice for Engineering Drawings;
- MIL-DTL-31000: Detail Specification Technical Data Package;
- MIL-PRF-28000: Digital Representation for Communication of Product Data;
- MIL-PRF-28002: Requirements for Raster Graphics Representation in Binary Format;
- MIL-PRF-28003: Digital Representation for Communication of Illustration Data.

Definition of CM

Configuration Management is not an easy thing to define. The many elements of this discipline, management methodology and area of professional expertise cannot be accurately and thoroughly communicated in a few words as can many job descriptions. For example, if you ask a mechanical engineer what they do for a living, they can say: 'I design mechanical objects'. It's not quite that simple to define your job as a CM practitioner.

So, in order to be able to explain to the casual observer what CM is, I use the following definition:

CM is a management discipline used to capture and control product data.

You can also say that:

CM is a way to get the right data to the right people at the right time.

I believe that this 'short' definition is accurate and a lot easier to use during casual conversations than the more detailed definitions which I will present to you in the following paragraphs. You may find, however, that these 'traditional' definitions are more appropriate when communicating with your peers or with someone who really needs to know the details of the CM job.

MIL STD 973

As applied to configuration items, a discipline applying technical and administrative direction and surveillance over the life cycle of items to:

- Identify and document the functional and physical characteristics of configuration items.
- Control changes to configuration items and their related documentation.
- Record and report information needed to manage configuration items effectively, including the status of proposed changes and implementation status of approved changes.
- Audit configuration items to verify conformance to specifications, drawings, interface control documents, and other contract requirements.

As applied to digital data files, the application of selected configuration identification and configuration status accounting principles to:

- Uniquely identify the digital data files, including versions of the files and their status (e.g., working, released, submitted, approved).
- Record and report information needed to manage the data files effectively, including the status of updated versions of the files.

How would you like to articulate that job definition to someone who just asked you what you do for a living?

EIA 649

This made it a little easier by providing the following definition of CM:

A management process for establishing and maintaining consistency of a product's performance, functional and physical attributes with respect to its requirements, design and operational information throughout its life.

That's it for CM definitions. I'll leave it up to you to choose the one you think most appropriate for any particular situation.

Configuration Identification

Configuration Identification deals with the subject of capturing and documenting designs at appropriate stages in the life cycle of the product. It describes the process of identifying the design and the hardware and software. *Note*: Always think in terms of both the design and the physical product, whether hardware or software or a combination of the two. The reason for this will become clear as we proceed.

Baselines

One major improvement in the content of MIL STD 973 and EIA 649 was the treatment of the subject of baselines. Experienced CMers know the importance of baselines. This knowledge, however, was usually attained as a result of battles, both public and private. The very notion of baselines was often foreign to businesses. Designs were often generated by electrical and mechanical designers and draftspersons with the final approved versions being tossed into the product design 'pool'. Nothing even remotely resembling a baseline was captured or tracked during the life cycle of the product.

In those cases where baselines have been identified, captured and tracked, CM activities are, for the most part, being conducted using a paper trail. This is not to say that paper is not an effective way of dealing with baselines and the change process, only cumbersome and labour intensive.

The main idea with baselines is to first identify the list of new and/or modified designs that, when assembled together, become the final product design and to capture the related drawings, specifications and digital design files at a predetermined time. From that point on, proposed changes to these designs are handled in a different, more formally controlled, manner.

Today

Baseline capture today involves the transfer of ownership of the drawings and specifications which define the product to an independent organization which then permits updates to those documents only after formal approval of proposed changes by either internal Configuration Control Boards (CCB) or external CCB or both.

The best practice, as described in Chapter 2, is the establishment of states and transition events based upon design reviews. Development baselines are established, and the corresponding documentation is captured at the transition to formal internal control, i.e., at Design review 3.

The drawings are 'issued' and put in the documentation control 'vault'. These drawings are only removed from the vault upon approval of formally processed changes. Upon incorporation of the approved change, the drawing is returned to the vault.

Tomorrow

'Tomorrow', as used throughout this chapter, is the transitional period of time (six months to a year or two) wherein we incrementally implement the functionality of our PDM tool and systematically integrate it into our business environment. It is the only way we can safely reach full implementation of our PDM without having elements of CM 'drop in the crack', thereby losing control of our development and production activities. These incremental steps are the 'twists and turns' that I mentioned in the Introduction. They will include the workaround procedures necessary to the phased functionality enhancements which we will utilize to achieve our long-term goals.

Also, remember that our future state of automated CM is going to represent a 'culture shock' for many. A significant readjustment of people's mindsets must occur. This process takes time.

Before I get too far into what we are going to do during this transition period, I should give you a brief overview of what a PDM is, what it does and how it does it. A detailed description of the PDM tool is presented in Chapter 9, Automated CM.

I'm using the term PDM to refer to a system used to capture and control product data in an electronic format and to make those data available to all who need to use or observe them. This includes capture and control of design data (native CAD and neutral files), quality assurance data, manufacturing data, and any other data that would be needed to design, test, build, modify, deliver and maintain a product (either hardware or software).

The heart of this system is a software Configuration and Data Management (CM/DM) tool which, in conjunction with third-party tools (image file view/ markup tools, file format conversion tools), provides the functionality required to implement and integrate the PDM system within any given business, educational or other environment.

The PDM can readily be expanded to include the capture and control of other data, such as internal procedures, contract data, proposals, etc. It would then be referred to as an Enterprise Data Manager (EDM).

In the past, designs were captured on mylar and paper drawings. In the general case, designs being generated now, and in the future, are, and will be, generated on Computer Aided Design (CAD) systems. The design disclosure master is no longer the paper drawing. Today's design master consists of the collection of digital design files and computer models, including part and symbol libraries, associations, linkages, etc., which define the product design. The need therefore arises to capture and control these digital design files in order to capture and control design baselines and to be in a position to incorporate approved design changes now and in the future. The PDM tool performs these functions.

The methodology described herein is in response to 'what is real today'. The 'what', 'how' and 'when' of future PDM configurations and procedures

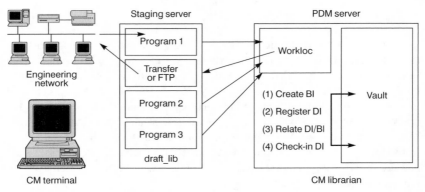

Figure 4.1 PDM staging areas

will be driven by a combination of customer, technology, cost, and best practice considerations.

The first step towards the implementation of a PDM system is the establishment of a staging area for the transfer of design files. This staging area must be located on a computer network that is accessible by both the engineering/drafting community and the CM organization (see Figure 4.1). A staging area named 'draft_lib' has been created in this illustration to demonstrate the use of staging areas. Design files are moved to and from this staging area by engineers and draftspersons and/or PDM 'librarians'. These files are retrieved from the staging areas by CM personnel and inducted into the PDM vault. This 'check-in' procedure varies among PDM tools.

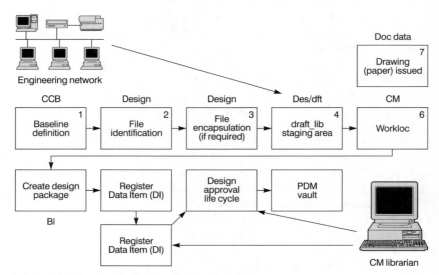

Figure 4.2 Electronic design file baseline capture

Baseline capture of electronic design files is initially conducted in parallel with baseline capture of paper drawings. The following list refers to the referenced activity numbers in the flow chart presented in Figure 4.2.

1 The first step in the baseline capture process is the identification of the drawings and specifications that comprise the design disclosure documentation package for the product as in the case with baseline capture of paper drawings. This list of drawings and specifications can be obtained from Drafting Planning Sheets, Family Trees, discussions with design personnel, and design review documentation listings.

2 The next step is the identification of electronic files that make up these designs. Simple mechanical designs may be made up of two-dimensional CAD files, whereas complex electrical and mechanical CAD designs may be developed from sophisticated models with associations and linkages, symbol and part libraries, schematics, simulations, etc. The appropriate files must be identified and defined, according to the design tool on which they were created.

3 Two-dimensional 'flat' files are captured without encapsulation. Models, depending upon the tool on which they were created, may be three dimensional and must receive special treatment. An encapsulation process has been devised to serve as a workaround until interface modules are available to achieve tight integration between the design tools and the PDM. This encapsulation process is intended to preserve the associations and linkages inherent in the model structure. Pointers to and from libraries and parts are preserved, and all files necessary for future reproduction and update of the design are captured and properly related.

4 The procedures for encapsulation and for moving the files to the draft_lib staging area vary according to the design tool used to generate the design. In all cases, however, these files must represent the design models, libraries, drawing files needed to reproduce the design in the future for incorporation of approved changes and/or re-use of the design on other products. The following paragraphs define the current approach for this process, using typical mechanical and electrical model design tools as examples.

When the mechanical design electronic file baseline is initially captured during the transition to the design internal state, i.e., Work in Process (WIP), only the drawing files are captured. Today's mechanical design tools generate both models and flat files that represent the equivalent of the old drawing. The model remains under the control of the design lead or drafting, depending upon individual programme roles and responsibilities and is updated by engineering and/or drafting. The drawing files are moved to the draft_lib staging area by engineering or drafting. They will serve as a 'snapshot' of the design at this time.

During this phase, when the electronic drawing is captured, the filename, version and revision number will be noted on the drawing (for both the model and the drawing files). For each change to the model which causes a new version, a red-lined drawing may also be created to document the change.

When the mechanical design electronic baseline is captured or updated during the transition to the formal internal state (development baseline), the design model and all its associated linkages and libraries shall be encapsulated and moved to the draft_lib staging area by engineering or drafting.

The electrical design database is initially captured during the transition to the formal internal state (development baseline). All native CAD design files are encapsulated and moved to the draft_lib staging area. Electrical designs are complex by nature and require a structured discipline to identify and locate files generated during the design processes. Encapsulation of related files serves this purpose well.

Files are categorized and accumulated according to the type of design (ASIC, power, module) and are then encapsulated. Neutral (hpgl, raster, etc.) drawing files for electrical designs (schematics, layouts, parts lists) are also captured at this time.

When two-dimensional mechanical (native CAD) design files are captured, they are moved directly to the draft_lib staging area without encapsulation. Neutral file versions of these native CAD files are generated at this time to be used for viewing and CALS-type electronic distributions.

5 Project folders are created within the draft_lib staging area on the staging server, as shown in Figure 4.1, by the CM organization. The design files (neutral files or encapsulated models) are moved into these project folders by engineering or drafting.

6 The files are then moved from the project folder(s) on the staging server to the workloc directory on the server which contains the PDM software and the controlled storage vaults. These files are submitted to an 'initial design approval' life cycle (controlled by the PDM) and then checked into the PDM vault.

7 At this point, in a parallel activity, the paper drawing is signed off internally and issued through the documentation distribution centre.

8 The next evolution of this methodology eliminates the redundant paper process (see Figure 4.3).

Future

In the future, when interface modules are available to tightly integrate the PDM with the model design tools (electrical and mechanical), the design engineer and/or design draftsperson will be able to initiate their design by simply signing on to the design network through the PDM and launching the desired design tool through the PDM. The designers will then work on their design throughout the day, and at the close of the business day, they will sign off the design tool, and the PDM will deposit the data

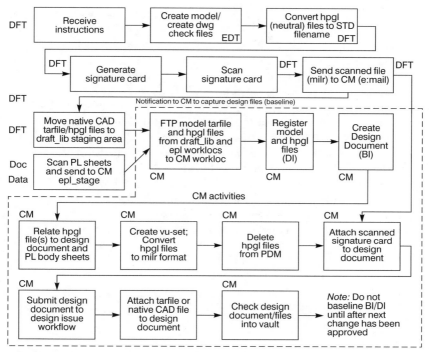

Figure 4.3 Baseline capture – paper eliminated

created during the course of the day into the designer's personal work location.

This process will be repeated until Design review 2 is completed. From then on, the 'workflow manager' application in the PDM will move the design files to a logical vault, and future changes will only be permitted after on-line approval from the designer's supervisor. Details of this process are presented below in the discussion on configuration control.

The point here is that baselines will be captured at design reviews by the PDM. Baselines will not have to be manually captured by a CM person and the corresponding files moved to and from staging areas. These data will be entered once into the PDM database and updated by engineering personnel under the control of the PDM workflow manager.

Enter data once – use many times! You can imagine the savings in time and money.

Configuration Items

One of the more confusing topics of CM is Configuration Items. Although the main thrust of this book is to describe a practical, state-of-the-art CM process with a minimum of nit-picking, I think it is appropriate to clarify the topic of Configuration Items.

Let's try to think about Configuration Items in the same manner in which we think about vehicles. There are many types of vehicles, e.g., cars, planes, boats. They all provide basically the same thing – transportation. However, there are clear differences between them and what they do. This is also true of Configuration Items but the problem here is that the different kinds of Configuration Items are called the same thing – Configuration Items.

People tend to put Configuration Items into three categories. I will explain these categories (and apply my own naming conventions to them) because you probably will run up against them, and you should be prepared to recognize them and give them the proper amount of consideration. However, after describing the current state of confusion, I will provide you with a practical definition of a Configuration Item which will 'do the job', nicely.

I will identify these multiple definitions, using my own terminology*, as follows:

- Technical Configuration Items (TCI)*
- Contractual Configuration Items (CCI)*
- Serialized Configuration Items (SCI)*

Technical Configuration Items (TCI)

TCI are those items which a business wants to control and follow for configuration management purposes, i.e., baseline, track changes to, report on and audit. Basically, every designed part or, in other words, every part with a drawing number assigned by the design cognizant company, is a TCI.

Contractual Configuration Items (CCI)

CCI are line items specified in a contract, e.g., guidance system, weapon system or subsystem, automobile, spares. They are made up of many TCI. This use of the terminology Contractual Configuration Items is infrequent, but very real when one is dealing with contractual delivery requirements.

Serialized Configuration Items (SCI)

SCIs are the most misunderstood of the lot. These are the subset of parts and/ or assemblies within the top assembly that have been selected to be serialized and tracked throughout their life cycle (or a portion of it). See paragraph on serialization, below. Actually, this category of items should not be referred to as Configuration Items at all, unless, by reason of some other characteristic, they can be defined as Configuration Items.

Now for the *Practical CM* version of Configuration Item. *Note*: I have had a considerable amount of help in formulating this definition, simple as it seems. CM professionals from government and industry have generously contributed a wide range of philosophical and practical intelligence so that this concise definition of Configuration Item can be articulated.

A Configuration Item is a physical thing that we want to identify and control throughout its life cycle. It may be a single part or line of software code or an assembly of parts or an entire software program or any combination thereof.

Thus defined, this thing we call a Configuration Item may be used to satisfy any or all of the three categories specified above.

Good luck with your CIs.

Serialization

The serialization issue is a trade-off of the availability of traceability information versus cost. If you want to, or need to, worry about such things as technical risk, vendor availability, spares (or component) traceability, serializing parts and/or assemblies may be a good idea. Serialized assemblies also make good targets for production cut-in of approved changes.

But all of this costs money. The items to be serialized must be selected early in the programme and, possibly, approved by the customer. Then a note must be put on the drawing (and on manufacturing planning) that says, 'serialize me'. Finally, as the parts and assemblies are manufactured, the serialization data must be captured and processed. 'As-built' (or verification) reports must be generated and distributed. Of course, in our future world, no reports are necessary. Users of these data will just browse our databases to extract the needed information.

The key, therefore, is to think twice about the items you are going to select for serialization. Think in terms of cost versus traceability. How much do you really need to know about what item is where in the future? As a rule of thumb, the serialized items used to be about ten per cent of the total parts in a military product. Now, the number is closer to five per cent. Usually, technically risky items are selected but if solid engineering methodologies are followed, this category of risk is minimized.

Part marking

Part marking is broken down into two categories: prototype part marking and production (or tactical) part marking.

Prototype part marking

The main reason for marking prototypes is to acquire 'snapshots' in time of the state of the developing prototype hardware and to correlate those snapshots with the actual design (on paper or in a database) at that specific time. In other words, as we move towards the final design, we find and correct problems with both the design and the prototype hardware and software. It is essential, therefore, to record changes to both and to tie them together for traceability purposes. If it is not mandatory to know the equivalent design status of the hardware until a certain time, then incremental prototype marking requirements may be relaxed or waived altogether.

As we shall see in the section on Change control, changes to the design in the design internal state may be recorded via red-lined drawings (prior to design review 3). These changes may also be reflected in the design database on an incremental basis or they may be incorporated all at once just prior to design review 3. In either case the drawings should be marked with numeric revisions, i.e., 1, 2, 3, . . ., etc., in order to distinguish between the WIP design revision levels and the post-issue alphabetic revision levels, i.e., A, B, C, . . ., etc.

The prototype hardware should also be marked with the corresponding revision number. This marking may be of a temporary nature such as hand marking or tagging but some form of marking should be affixed to the prototype hardware in order to identify the state of the design which it represents.

Production hardware marking

Production hardware must be formally marked per MIL STD 130 (tactical military hardware) or via formal company marking standards (commercial) with the drawing number, serial number (if applicable) and other markings defined by contractual requirements.

Documentation levels

There are two major categories of documentation:

1 Design Approval Documentation;
2 Design Disclosure Documentation.

Design Approval Documentation

Design Approval Documentation consists of all the initial and developing design information generated along the path to the final design. It includes but is not limited to sketches, layouts, pre-issue drawings, parts lists, wire lists, engineering notes, design review minutes and action items, preliminary design specifications, test plans, meeting minutes and view graphs, and any other documentation which defines the pre-approved design.

Design Approval Documentation is generally documented informally. This means Level 1 (DOD STD 100 on military contracts) or developmental documentation (per MIL STD 31000).

Design Disclosure Documentation

Design Disclosure Documentation is just what it sounds like, i.e., the formal disclosure of the final, approved design. *Note*: When I use the terminology 'approved design' throughout this book, I am referring to a design which has been subjected to a thorough, formal review and approval process both internally and with the customer, if that is what is specified in the contract.

The Design Disclosure Documentation must be a 'stand-alone' package. This means that another bidder, supplier or contractor must be able to accurately bid and perform on a contract for this product by utilizing this documentation without recourse to the initial designer. In military parlance, this documentation package is DOD STD 100, Level 3.

Well, what about Level 2? Level 2 is Level 3 to start with. Then, as we tailor and negotiate modifications to contractual requirements (remember Chapter 3?), Level 3 becomes Level 2, tailored! Simple as that.

Technical Data Package

The Technical Data Package (TDP) consists of parts of the Design Approval Documentation (determined by the customer) and the entire Design Disclosure Documentation. It is sometimes a deliverable item under a contract and sometimes not. The TDP is broken down into four (self explanatory) categories:

1 Deliverable hardware and software documentation;
2 Non-deliverable hardware and software documentation;
3 Tactical hardware and software documentation;
4 Non-tactical hardware and software documentation.

Configuration Control

Configuration Control deals with the subject of controlling changes to baselines throughout the life cycle of the product. This section presents the reader with 'best practices' for establishing and maintaining efficient change processes. It also provides guidelines and rules for assuring that changes to the design are incorporated into design documentation in a timely manner, that these changes are cut into production as planned, and that obsolete or unacceptable parts and assemblies are removed from service, whether to be upgraded or scrapped.

Today

In keeping with our scenario of discussing today's processes first, then moving on to the transition period as we move ahead on our journey to excellence in CM, let us examine the multi-faceted process for change control (a subset of Configuration Control) practised in today's military and commercial world.

Configuration Control Boards

Configuration Control Boards (CCB) are formed at the point in the development cycle where the design(s) are transitioned from informal control (Design review 3) to formal internal control. Refer to Chapter 2, Figure 2.1.

The composition of the CCB(s) may vary from programme to programme and from internal to external CCBs. The members of the CCB are selected by the programme manager or an individual of equivalent authority. There should be, at a minimum, representatives from engineering (systems and design), manufacturing, quality assurance, configuration management, and product or field support. Additional representation from producibility engineering, components engineering, reliability engineering, plus human factors and safety may be called upon to support this activity as required.

On the other hand, if you have only one or two engineers in your organization, they plus a management representative may comprise the CCB. That's OK, too.

The idea is that the appropriate people review proposed changes to the design. This includes a review of these changes for technical merit and impact, related production impact, cost and schedule impact, plus just good common sense.

Approval by the customer may or may not be required, depending upon the programme, the point at which the change is proposed during the development or production cycle, whether or not the design is for a military product, and the class of the proposed change.

Problem resolution

The first step in the change control process is problem identification and resolution. Problems are brought to the attention of key individuals in the business by designers, users, builders and maintainers of the product. They are also sometimes identified by the customer. The forum for the introduction of these problems may be formal, i.e., Problem Resolution Board (PRB) meetings, Corrective Action Committee (CAC) meetings or informal contact with programme managers, project engineers or other individuals within the company.

As in the case of the CCB, the important point to remember is that the problem receives proper attention by the appropriate people.

Problems may be solved directly as soon as they are identified or they may be referred to a design investigation, which may be short or long in duration.

In either situation, the resolution of the problem may require a design change, in which case the action is normally referred to the CCB for consideration of impacts to the programme or project. The recommendation from the individuals charged with the problem resolution activity may consider all impacts but sometimes Problem Resolution Boards neglect or are unable to determine the impacts on all functions and activities of the company.

Change proposal

A formal process should be established and maintained (audited) to recommend, review, approve or disapprove, process and implement (in both design and hardware/software) the design change.

The format of the vehicle(s) utilized to process the change vary from military to commercial products, and also among the various military services.

A standard Engineering Change Proposal (ECP) form is provided in MIL STD 973, as are standard forms for proposed Specification Change Notices (SCN), Notices Of Revision (NOR), and other specific purpose forms. Also, internal company Change Notices (CN) are used to capture the 'from–to' information from ECPs and NORs plus add effectivity information, i.e., when to cut the approved change into production.

I will not include detailed instruction in this book on those forms or on the procedures used to fill them out or process them because we are dealing here with recommended 'best practices' for today, tomorrow and the future. The goal is to identify the steps we must take to lead us into a world without forms and unnecessary paperwork. I feel quite comfortable in stating that it is the government's and industry's unanimous intention to do away with forms altogether and to replace them with computer screens which will accommodate the entry of data (attribute information), either automatically or manually.

We will concentrate instead on what you need to know to move from the labour-intensive morass of manual change form circulation, review and approval. Next, we will take the steps necessary to improve the efficiency of the drawing update process. Finally, we will dispose of the baggage of wasteful paper distribution systems, to say nothing of lengthy paper drawing update practices and last-minute incorporation into the product hardware and/ or software.

However, to keep this all in perspective, we must first identify the current steps in the change proposal and approval process in order to appreciate our progress in our journey towards excellence in CM.

Change approval

The change approval process usually starts with the design engineer or project engineer recommending the change to the CCB. The CM representative to the

CCB will have reviewed the change proposal paperwork prior to the meeting for format and have it ready for CCB member review and discussion at the meeting. In some cases, the change proposal paperwork is distributed to the CCB members prior to the meeting to allow for pre-meeting review.

During the meeting, the members of the CCB will ask questions, each according to their area of expertise. A discussion often ensues, and spontaneous questions and answers are posed to ferret out the information necessary to the thorough challenging of the proposed change and the careful and in-depth consideration of its probable and possible impacts on all related functions and activities. *Note*: We will deal with this interactive facet of the CCB meeting in our discussion of on-line CCBs. After all, we don't want to sacrifice the quality of decision making for the expediency of change approval cycle-time reduction.

Once the decision has been reached to approve the change, the CCB representatives charged with signing the change vehicle (internal or external change proposal or change notice) will affix their names to the paper and move on to the next proposed change. If the change is disapproved, this fact will be entered into an appropriate database and the change proposal paper will be filed. The approved change paper then moves on to the next step in the process.

I can think of quite a few good things to say about the CCB meeting itself, and the quality of information captured and considered there if the meetings are conducted properly. Of course, as is the case for all meetings, good or bad things can happen. Usually the results are somewhere in between. It all depends upon meeting discipline, meeting procedures and the effectiveness of the moderator or chairperson in charge of the meeting.

We will be very careful not to lose any of this 'good stuff' when we go on-line with our CCBs.

Design update

I cannot make the same observation about the design update and distribution processes that I did about CCB meetings. Many processes in today's CM world are inefficient, labour-intensive, boring, tedious, and just plain ugly but the wastefulness involved in the update of drawings and the distribution of approved change paper and updated drawings is, almost without exception, abysmal.

Even the people engaged in these thankless tasks readily admit this to be the truth. They work just as hard as anybody else and are just as tired at the end of the day but the perception of the inefficiency of their busy work far outshadows the perceived added value of their contribution to the design control process.

Again a clarification is in order. We would not be able to control our change processes and meet our contractual commitments without these people and the 'difficult-to-see' fruits of their labours. However, there is a better way to get the job done.

After receiving approval by the CCB, the change paper becomes the vehicle for the update of the design drawings and/or design databases. The step-by-step procedure for the movement of the change paper is described below in the section about transitioning to the world of PDMs. Suffice it to be said that many steps are involved in the process – necessary today but not necessary in the future.

In addition to the update of the design media, the change paper itself must be distributed to those affected by the change, i.e., engineering, manufacturing, quality assurance, sourcing (purchasing), field support, field sites, vendors, subcontractors, suppliers, customers, and anyone else who needs to get into the act.

Paper copies of the updated design drawings also follow this circuitous route. When finally all is said and done weeks or months may have passed. Much water (and money) has flowed under the bridge.

Change implementation

Changes are cut into production according to the information contained in the effectivity section of the internal change paper. The Production Control representative to the CCB, in conjunction with the Manufacturing and Quality Assurance members of the CCB, assigns specific dispositions for various categories of materials. This activity includes both engineering prototypes and production units. The following are examples of the categories of materials to be considered:

- Material on order from vendors and suppliers;
- Material that has been received but has not yet started fabrication or assembly;
- Material in process on the shop floor;
- Material that has been partially or completely fabricated, assembled and tested;
- Material that has been delivered to the customer.

Possible dispositions include but are not limited to:

- Modifying;
- Scrapping;
- Doing nothing (Minor (previously Class 2) changes only).

The approved change may or may not be incorporated into these materials, at the discretion of the members of the CCB.

Additionally, for Major (previously Class 1) changes, all stock on hand or delivered to the customer must be purged from the system and replaced by the modified product. This is an excellent example of the need for good traceability records, i.e., a good Status Accounting database. (See Chapter 6 for an explanation of Major and Minor changes.)

Tomorrow

The operative thought here is that we don't want CM to slow down design, development or production. We want to make the process easier and faster. Therefore, we don't want to interrupt what is already working. We want to gradually supplement and finally replace our antiquated methods.

So, as with baseline capture, we will integrate the electronic change process with our existing paper change process by incrementally increasing the functionality and flexibility of our PDM tool.

The following list refers to the referenced activity numbers in the flow chart represented in Figure 4.4:

1 The first step in the transitional change control process is CCB approval of a proposed change. *Note*: Let's call the change 'vehicle' associated with the change proposal and approval process a Change Notice (CN). This CN will be the equivalent of a folder which will contain all forms, internal and external, associated with a proposed change, such as the ECP, NOR and the internal change paper used on various programmes. We might as well start simplifying things at this point, since, when we get right down to it, we will be entering data on a computer screen in the future. This screen will provide fields that satisfy the necessary change proposal information for any situation.

   We have a computer screen for the CN and, reporting to it, other screens for ECP, NOR, waivers, deviations, problem reports, Change Notice Disposition (effectivity), drafting/engineering signature sheets, plus related drawing files and the 'from–to' composite or 'compare' images or red-lined drawings (represented by markup overlay files

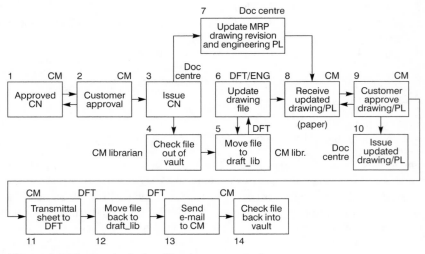

Figure 4.4 Electronic design file change procedure

created by your PDM 'view' tool). In other words, we will enter the information and files into the PDM that will provide everything we need to know about the change that we are about to process, both for today's immediate requirements and for traceability information in the future.

2 Upon receipt of approval by the CCB, the CN is provided to the customer for their review and approval by the CM representative to the CCB. This may be an iterative process, involving coordination of customer requests and engineering responses to those requests until the customer is ready to approve the CN.

3 The approved CN is delivered to the Documentation Centre by CM. The Documentation Distribution Centre issues (puts a date on) the CN and distributes copies internally and externally. *Note*: Drafting should receive their copy of the approved CN the following day. CM also notifies the CM librarian that a CN has been approved, so that the CM librarian can make the appropriate files available for update.

4 The CM librarian checks the design files needed to incorporate the design change out of the controlled library vault and places them in the PDM staging area.

5 The CM librarian then moves these files to the engineering server 'draft_ lib' staging area so that they can be retrieved by the appropriate draftsperson or design engineer for update.

6 The draftsperson or design engineer retrieves the files from the 'draft_lib' staging area and proceeds to update the file(s) as required to incorporate the approved changes specified in the CN.

7 In a parallel activity, your Documentation Centre updates your Material Resource Planning system (MRP) (this varies from company to company) for CN approval date and drawing revision level.

8 CM receives a newly plotted copy (paper/mylar) of the updated drawing, plus a transmittal sheet from drafting and provides this documentation to the customer for review and approval. *Note*: This may also be an iterative process, as in Step 2.

9 The customer approves the updated paper copy of the drawing(s). CM provides the updated and approved documentation to the Documentation Centre for issue.

10 The Documentation Centre issues and distributes the updated documentation.

11 CM returns the documentation transmittal sheet to drafting.

12 Drafting moves the updated and encapsulated files back to the draft_lib staging area.

13 Drafting sends an e-mail message (with the appropriate signature card raster image files attached) to the CM librarian as a notification that the updated design files are available for check-in to the PDM vault.

14 The hardware CM librarian checks the updated files back into the PDM vault.

Note: The design change proposal and approval procedure detailed above represents the first step in the transition from our old paper change process to an electronic data management scenario. It is a step in the right direction but it still involves an overlapping, redundant and expensive paper process. This process should only be used until your personnel are comfortable with the electronic data control part of the process. It will then be time to move on to the next step towards automated CM.

This next step disposes of the use of paper altogether for the design change proposal submittal and approval process. Figure 4.5 presents the recommended procedure for flat (raster) files. Figure 4.6 presents the procedure for design models and native CAD files in general.

This concludes my advice on how to start the process of transitioning your present paper change control system to an electronic PDM-managed system for this specific area of CM. Implementation of the required PDM functionality and integration of this process into your business programmes is presented in Chapter 9.

Companies vary widely in the way in which they practice change control today. That's OK, though. By implementing a PDM approach to change control in parallel with your current techniques, you will find the most suitable path to your future state of automated CM.

Keep in mind that you will make mistakes. This is a complicated process. You should garner valuable 'lessons – learned' from each mistake. Refer to Appendix D for several examples of 'lessons learned' by myself and the

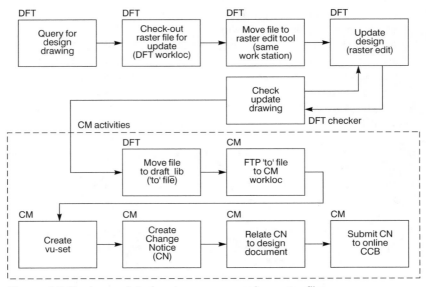

Figure 4.5 Design update for change proposal – raster files

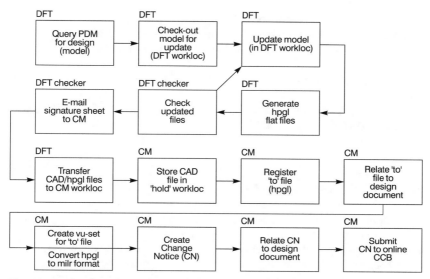

Figure 4.6 Prepare change package – design models

companies for which I have worked. The sure road to success is guaranteed by persistence, patience and the willingness to take a few arrows in the back.

After all, that's what pioneers do!

Future

In the future, changes to the design files will be accomplished based upon electronic authorization from your design lead or internal and external on-line CCBs. PDM life cycle managers (functional modules) will move design files from one logical vault to another logical vault or, in some cases, physical vaults, depending upon the PDM functionality integrated into your business systems.

Designers will update files (baselines) that have been captured and maintained by the PDM throughout the life cycle of the product. Your designers plus drafting, manufacturing, project, and CM personnel will provide the necessary change approval information to the PDM via input screens. These screens will accompany the 'from–to' data input to the PDM by the designers and draftspersons.

Simultaneous approval of the proposed change by several members of your CCB(s) will be incorporated into the on-line CCB functionality if your business process warrants it.

Changes will be cut into production much faster because of immediate notification to production control and manufacturing personnel of approved changes. Designs will be controlled, tracked and available at all times to multiple users and to your customers.

As in the capture of baselines, we will enter the data once and use it many times.

Chapter 9 will bring the whole picture into focus from the standpoint of automated CM and the PDM.

Configuration Status Accounting

Configuration Status Accounting (CSA) is the CM activity that provides information about the current status of approved baselines and the progress and status of proposed and approved changes to the design. CSA also covers the activity of non-conforming material status reporting, i.e., as recorded and processed on the form, Request for Waiver/Deviation (RFW/RFD), including the resulting corrective action. *Note*: This form is headed for the scrap heap along with the change forms mentioned above. The information must still be captured but on screen or directly into a database, not on paper.

I will break down CSA activities into three distinct but overlapping areas:

1 CSA data capture;
2 CSA data processing;
3 CSA data reporting.

Great news! The very nature of these activities will enable their transition to automated CM via a PDM as easily as a horse will find its way back to the barn.

Today

Today, the methods utilized to capture, process and report on the various types of CSA data are a combination of manual and automated activities. I will surely fail if I try to list them all here but a reasonable sampling is in order and do-able. Besides, when I describe the direction this area of CM is going to take, you will see that a thorough understanding of all of today's processes is unnecessary.

CSA data capture

There are several types of CSA data. There are the contractually required deliverable data, and there are the data that we capture to track the progress of proposed changes. There are the data that we record to determine whether or not approved changes are incorporated into the product (hardware and software) as defined by the CN. There are the data that capture the dates that events occur, i.e., approvals, disapprovals and who gets what when. There are the data that classify and pigeon-hole the types of change and the data that categorize the reason for the change – not to lay blame, of course, but to provide 'metrics' for the greater good of mankind. Pardon me while I take a deep breath.

In other words, there are data that are required, and there are data that are 'nice to have'. Both categories of data will be provided at a small fraction of the current cost by our PDM.

By the way, in case you wondered, even though it seems clumsy to say 'data are' as opposed to 'data is', are is correct. Data are plural. Datum is singular. Don't worry. We won't use the term 'datum' in this book.

CSA data are currently being entered into various flavours of PC databases (usually the 'change trail' data) and even into paper logbooks. Some data are automatically recorded by CSA databases as events occur in the capture and control of designs. Some data are entered into more than one form of record and then transferred onto another form of record. This is a good example of redundant work that will be eliminated by your new PDM system.

CSA data processing

The processing of CSA data includes the transformation, where required, of the data from its 'captured' format into the format required by your business's or your customer's computer or manual systems. These data will be utilized for analysis and/or delivery of the CSA data to your customer in order to satisfy your contractual requirements.

This transformation may be simple keyboard entry of data into a database, or it may involve a combination of manual data recording and automated data processing techniques.

Chapter 9 will provide a detailed explanation of the way in which your PDM system will capture, process and make available to all users those data required for each of your internal and external functions plus satisfy your customer's needs, so that everyone has the necessary information available to do their jobs.

The analysis of CSA data may amount to a visually superficial examination of records to determine that the correct type, format and quantity of data were recorded. On the other hand, a detailed analysis of CSA data may be conducted to determine ways to reduce cycle times and to identify bottlenecks in CM, development and manufacturing processes.

CSA databases may be automated to 'flag' past or near due CDRL deliveries. Manufacturing operations generally use CSA data to initiate material orders to prepare for production cut-ins. The list goes on and on.

Suffice it to say that CSA data are widely used and can be very useful.

CSA data reporting

CSA data reporting today is mainly via delivery of paper reports, PC and mainframe database printouts, and aperture cards.

This method has been documented in Military Standards, Data Item Definition (DID) forms and recorded in a multitude of internal company procedures and operation manuals.

It works but it takes a lot of time, is labour intensive and slow. There is a better way.

Tomorrow

The first step along our road to excellence in CSA is the transition to delivery of data in electronic format. This can mean delivery via magnetic media, i.e., tape, CD ROM, or delivery over the 'wire' in CALS format such as IGES (Initial Graphics Exchange Specification) or raster or any of the many 'neutral' file formats in use today, e.g., PDF.

CALS is currently defined as Continuous Acquisition and Life Cycle Support. It has had other definitions but the idea is the definition of a standard set of file formats and compositions that one business can use to provide information to another business, or its customers and suppliers. This method of communication can apply to individuals as well as businesses. IGES and raster are just two of the many acceptable formats used to deliver data under the CALS umbrella. PDF is fast becoming a 'de facto' CALS standard for text data.

The subject of data formats is a moving target as you might expect. As technologies advance and data processing speeds increase and as the demands for manipulation of data become more focused, the need for greater resolution and speed in the data transmission field becomes more pressing.

The methodology I will recommend for implementing data delivery utilizing our PDM system is straightforward.

The PDM maintains data in vaults and work locations. These data, in the big picture, represent the 'enterprise' data where the enterprise is your business or company. In other words, whatever you're looking for, it's in there.

The PDM data model defines the 'objects' which contain information and files. This information can be design data, contractual data, company procedure and process data, proposal data, or any type of data used in the business. The best method for storing data that may be updated or changed in the future is in the native CAD format. This is the format that is used by the software tool that created the data in the first place. Use these data for future changes.

At the time you capture the native files, you can use a software utility program to store another version of the same files in a separate object. We will call this our 'neutral' file format. This is the file that will be used to view the actual image of designs (equivalent to a drawing in the old world) and to make deliveries to customers and suppliers in the near term.

This utility will add 'header' information to the neutral file, such as drawing name, number, revision level, number of sheets, sheet size, CAGE (Commercial and Government Entity) code and anything else your contract calls for when you deliver data to your customer.

What we have now are files to be used to update our designs and data plus files for viewing 'drawings' of our designs. We also have data plus files for physical delivery to our customers and suppliers. Chapter 9 will provide the implementation details.

Future

Our final goal in our journey to excellence in CM is the true 'CITIS' environment. CITIS stands for Contractor Integrated Technical Information Services. It is the way our government intends to capture and control technical data resulting from contracts it awards in the future. Make no mistake. This will be the world of tomorrow. If you want to play in this ballpark, you will have to be CITIS compliant.

So what is CITIS? It can be thought of as 'the big database in the sky'. The idea is to enter data once and use it many times. No more delivery of data to anyone! The designer starts off by entering their design data and design files into the PDM system and then updating these data and files as the development process evolves. Orders will be placed with suppliers who will browse the database according to permissions provided by the PDM. Customers will check on design, test, material procurement and product fabrication and delivery status by browsing in this database and by viewing the design images, limited only by the security access permissions assigned by your company.

In short, everyone who now uses paper to gain access to information required to do their job will, in the future, go to the CITIS database to retrieve the information or files they need.

Enter data once – use many times! You will hear this phrase articulated often in the future. The concept is simple but extremely important.

Imagine the savings! No more paper or aperture card distribution. No more microfilming. No more CALS data and file deliveries (and that was a lot better than the paper deliveries). Instant access to data. Wow!

How do we get to this state of bliss? Well, it's relatively easy. If we make the near term transition as advised above by going down the CALS road, then it's a short step into the world of CITIS. The key is the network structure, including firewalls, which we must establish. Chapter 9 will provide the details for this ultimate realization of truly 'automated' CM.

Configuration Audits

There are three categories of Configuration Audits:

1 Functional Configuration Audits (FCA);
2 Physical Configuration Audits (PCA);
3 Configuration Verification Audits (CVA).

A brief recap of the explanation of these configuration audits, as presented in Chapter 2, will set the stage for our transition to production.

Today

Figure 4.7 delineates the activities to be performed during the FCA and defines the roles and responsibilities of the participants. Figure 4.8 provides similar information for the PCA. Note that the CM role in the FCA is more of a support function, and the CM role during PCA is that of leader, coordinator and doer.

Activity	Responsibility
1 FCA plan	Systems/CM
2 Design verification test plan*	Systems
3 Design qualification tests	Systems
4 Problem resolution	Transition team
5 Test reports	Systems
6 FCA report	CM
7 Test witness/verification	Customer (optional)
Establishes 'design works'	

*Includes the Requirements Traceability Matrix

Figure 4.7 FCA roles and responsibilities

Activity	Responsibility
1 PCA plan	CM
2 First Article Inspection	Manufacturing
3 PCA Rev/Ser # Data Capture	CM
4 Acceptance Test	Systems
5 PCA data processing	CM
6 Problem resolution	Transition team
7 PCA report	CM
8 PCA witness/verification	Customer (optional)
Establishes 'product baseline'	

Figure 4.8 PCA roles and responsibilities

Functional Configuration Audit

The FCA is conducted on the engineering prototype during the 'formal internal' state. The purpose of the FCA is to assure that tests have been conducted to verify that each requirement in the system level specification has been met by the design. If tests cannot be performed to verify a particular requirement, then a 'theoretical error analysis' must be performed to verify satisfactory compliance to the requirement. These tests are generally referred to as design evaluation and qualification tests. *Note*: The advent of the terminology FCA and PCA is relatively recent. It is mainly driven by military standards and used primarily on defence programmes. The purpose of these activities is universal, though, and should be an integral part of every development programme.

Physical Configuration Audit

The PCA is initiated upon successful completion of the FCA. During the PCA, the engineering drawings are proofed by direct comparison of drawing data or design data to the physical characteristics of the first production unit. This unit should be built to manufacturing planning that was created based upon the engineering drawing package.

Drawing measurements are verified against the actual hardware. All instructions, processes and technical data specified on the engineering drawings are also verified against the hardware. Historically, many businesses referred to this activity as the First Article Inspection (FAI).

The second major activity of the PCA is the capture of drawing revision and serial number data and comparison to 'as-defined' revision levels plus a verification that serial numbers have not been previously used.

The final activity of the PCA is the performance of an Acceptance Test (AT) to assure that the unit examined is physically and functionally the same as the unit that passed the FCA. This AT consists of a subset of the design evaluation and qualification tests performed as the basis for the FCA.

Upon resolution of any problems that were observed during the conduct of the FCA and PCA, a FCA/PCA certification form is signed by the customer, and the 'product baseline' is established.

Configuration Verification Audit

CVAs are conducted on each deliverable production item as a check on the appropriate cut-in of Class 1 engineering changes, a verification of the revision level of selected parts and assemblies and the recording of serial numbers.

At the beginning of the development effort, a list of recommended parts and assemblies to be verified is generated by the company or contractor doing business with the DoD and, in some cases, commercial companies. The purpose of this list is to identify which items should be verified as they are produced and which items are to be serialized.

It's all a matter of cost versus traceability. Information is recorded during the verification audits. It costs money to record, process and report this information. A determination must be made early during the development period about how much information is really needed versus what information is simply 'nice to have'.

The considerations during the generation of this list should include:

- Technical risk (new designs);
- Vendor instability;
- Spares (traceability);
- Matched sets or pairs;
- Production cut-in verification for Major (Class 1) changes.

Whatever the reason, if it is felt appropriate to capture serialization information on a part or a group of parts and/or assemblies, these data must be captured, processed and reported.

The method of capturing configuration verification data is usually a combination of recording the serial numbers during the assembly process via manual marking on the planning paperwork, electronic scanning of bar codes, or keyboard entry of the data after visual examination of the product.

Tomorrow

Configuration audits are by nature, integral to the business activities of planning, data capture and evaluation. The purpose of these audits is to verify that processes are followed and that the designs developed and tested are what they should be. CM audits are meant to prove that the product which has been designed and produced 'works' and that the producing company has a 'build package' that can be used to reliably produce additional copies of the product.

The benefits of the PDM system are most visible and its productivity most apparent in the design data capture, control and distribution activities of a programme. The benefits of the PDM system as far as CM audits are concerned is limited to the enhancement of the data capture and data processing activities. The planning and evaluation activities are, by nature, heavily human dependent and are only supplemented and enhanced by the visibility and data retrieval capabilities of the PDM.

With this in mind, let it suffice to say that the PDM system that we are going to put in place for the interim functionality of data capture and control will easily handle the demands of the CM audit activities. Chapter 9 will articulate the specific processes involved in this implementation as we travel on towards our goal of excellence in CM.

Future

The ultimate goal of the full implementation of CITIS functionality is, as we have said before, 'enter data once – use many times'! As we move towards realization of that goal, the CM activities of identification, control, status accounting, and audits will benefit from the inherent nature of the PDM system which we are going to implement, in a sensible fashion, in our businesses.

I could dwell on these benefits here but I think that your time is better served if I postpone the specifics of both the implementation details and the rewards to be derived from our new PDM system until my detailed coverage of the automated CM process in Chapter 9. So, stay tuned. The best is yet to come.

5
Transition to production

The transition to production begins during the formal internal state (see Figure 5.1). Following the issuing of the engineering drawings (design review 3) and the capture of the electronic design files (if you have the check-in/check-out functionality of a PDM in place in your operation), your MRP system is loaded with the engineering design information, and orders for production material can be placed.

In parallel with this activity, the Functional Configuration Audit (FCA) is initiated on your engineering prototype hardware and software. Admittedly, there is some degree of risk in placing orders for material before you have thoroughly proofed out your design via the FCA; however, due to the cost and schedule constraints present in today's business world, there is often no other way to complete the design and deliver the first production units on time.

As it turns out in practice, the engineering development methodology that should be followed allows for that 'expectation of goodness' in your design at the specific points we have chosen as your transition points from one level of

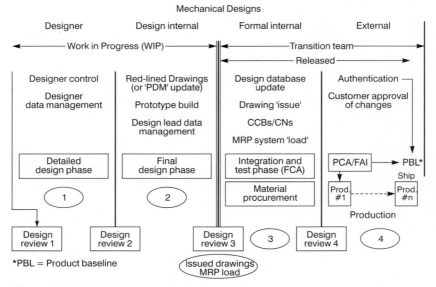

Figure 5.1 Transition team – programme phase

control to the next, i.e., your design reviews. Also, the added visibility provided by your PDM system plus its control functionality in the change approval process and faster cut-in of approved changes to production, reduces the risks even further.

The transition to production encompasses the entire Stage 3 shown in Figure 5.1 and continues on through the establishment of the product baseline in Stage 4. *Note*: I will be referring to Figure 5.1 often during this discussion, so please bookmark it for easy reference.

In Chapter 4, I followed the practice of first defining a CM activity and then discussing how that activity is performed in the business world of today. I then moved on to illustrate how we will improve upon the processes controlling that activity by installing a PDM system in an incremental fashion, until we have achieved a fully 'automated' CM process in the future.

In this and subsequent chapters, I will assume that you have established, or are well on your way to establishing, the fully automated CM process via full integration of a PDM in your business. Chapter 9 describes this system in detail and provides the planning and implementation instructions necessary to integrate the functionality of a PDM system within your business in a phased approach. Whereas it was helpful and necessary to show the step-by-step evolution to the desired state in Chapter 4 it would be of little or no value to dwell on obsolete methods at this point in our journey to excellence in CM. The increased visibility, control and reduced cycle times that are made possible by your new PDM system will be apparent from the material presented in Chapters 5 to 8.

Transition team

The establishment of a transition team is the first step in your preparations for the transition to production. The very terminology 'transition to production' screams out the need for such a group of people. After all, what are you really doing during this phase of your programme?

You are taking your design, which has been developed according to the requirements put forward in your system level specification, and you are preparing to order the necessary materials to fabricate production units of your engineering prototype.

Different people are going to be utilizing your design data. Different people are going to be building hardware and, maybe, software items. Different people are going to be performing inspections and measurements and conducting tests.

It therefore follows that it might not be such a bad idea to get these people to talk to each other. Let's make them meet regularly. Let's call them the 'transition team'.

Now, who should you have on your transition team? The following list represents the 'best case' scenario for representation on a transition team. You may have all or few of these individuals in your employ. The idea, though, is to utilize the expertise of those personnel who have the engineering knowledge and the manufacturing know-how to move the design from its development state to production with a minimum of hitches.

Before I list the functions that will provide your potential transition team members, I should point out that many of the functions shown will look familiar to you. They should. They are, in several cases, the same functions who provided representatives to the CCB(s). The interesting thing in this case, though, is that more often than not, there are different personnel involved.

This is not because the design took so long to develop that the original designers left, retired or were fired for incompetence. It is not because the CCB members had so many battles while considering proposed changes that they can't stand the sight of each other's faces any longer. It is merely because a different skill set is required for this task.

The transition team members must be intimately familiar with material procurement, and manufacturing processes. They must know exactly what has to be done and when it has to be done. They must know, in the greatest detail, what they need to do their jobs and when those requirements must be satisfied.

The engineering personnel must have hands-on intimacy with the engineering prototype. They ideally will be the ones who conducted the performance evaluation and qualification tests. They must be familiar with the 'lessons learned' during the test phase of the programme.

Similar requirements apply to quality assurance, sourcing, and project management personnel. In many cases, only the CM representative is the same individual. Of course, they know everything and can do anything, right?

As you can see, the people I am talking about here are the 'doers' of the detailed tasks. Not that the CCB representatives are not dedicated, capable people. They just do different 'stuff'. The CCB folks generally have a lot of experience in broad areas, are familiar with company policies and goals, and are the trusted custodians of the designs and the production processes. The transition team members are the 'hands-on' people. Of course, in any given organization they may be the same people. I am just recalling for you what I have seen to be the case during my years on the job.

I obviously have let the cat out of the bag as to the identity of most of the transition team members. However, to recap, the best practice would be to include a member from each of the following functional areas:

- Design engineering (mechanical, electrical, software and components);
- Systems engineering;
- Test engineering (if separate from systems engineering);
- Manufacturing planning and production control;
- Manufacturing assembly;
- Quality assurance;

- Sourcing or purchasing (same thing);
- Configuration management;
- Logistics;
- Customer or field support;
- Technical publications (user's manuals);
- Project engineer;
- Program manager;
- Program engineer (if you have one).

Some of these members may serve in an 'on call' basis. However, you should have a core team that meets regularly to make sure 'all bases are touched' as you head into production.

Transition team meetings

Transition team meetings should, preferably, be held at least once a week following the successful completion of design review 3. Additional 'splinter' meetings may also be appropriate when items are brought up at the transition team meetings that cannot be readily or immediately dealt with. These meetings don't have to last long – one hour is appropriate.

Meetings should continue to be held weekly until completion of the PCA and resolution of any and all discrepancies and closeout of action items assigned as a result of either the FCA or the PCA. At this point, the product baseline is established and production continues with a fully 'proofed' build package.

Action item list

It is extremely important to generate, follow up on, and maintain current, an action item list for all activities addressed, discussed and disposed of at the transition team meetings. Specific individuals shall be assigned specific action items to be completed by specific dates. Only when complete ownership is assigned and accepted can there be assurance that tasks will be faithfully executed on time.

The project or program engineer should be assigned the task of following up on action items and making sure that closure is effected on schedule.

Database verification

A verification check should be performed on the engineering, manufacturing and sourcing databases to ensure that production material orders are placed against the proper revision level of the engineering documentation and the manufacturing planning. This effort will require cooperation between several team members as will most of the transition team's activities.

This 'preventive maintenance' step performed at this point in the programme can save significant cost and schedule problems down the line.

Updates to these databases shall be implemented immediately upon identification of errors, and the corrections shall be communicated to all affected individuals and organizations.

Release to production

I believe that this is an appropriate place to clarify a misconception that is held throughout industry as to what is 'release' and what is 'issue'. In my CM methodology, we 'issue' drawings and then 'release' them to production.

The 'issue' activity captures the approved baseline, both paper and electronic, based upon successful completion of specific design reviews. Then our MRP system is updated with the engineering design data, which allows us to order parts. Future proposed changes are formally processed through our internal CCBs. Customer involvement in our change approval process is only after establishment of the product baseline, i.e., FCA/PCA completion. If we are dealing with a military programme, we obtain customer authentication at this point.

As you can see, we have delegated the approval and 'baseline capture' activities to a process we call *issue*. That means the design(s) are under CM control. We then *release* the designs to our internal manufacturing people or to our suppliers, subcontractors and vendors to procure material for and to fabricate the products.

Also, in regard to a schedule to do all this good stuff, we base our baseline capture (issue) upon design reviews, where a certain amount of 'goodness' in the design is expected at each design review. This has proved more reliable than committing to schedules that often slip. These design reviews are an integral part of our engineering development methodology. People pay attention to them. They should be conducted per an approved programme schedule, and thus make CM planning more reliable and dependable. People know their roles and responsibilities and know what and when something is expected of them.

Material procurement

Major vendors and suppliers should be invited in-house for reviews with the transition team. These 'up-front' meetings will serve to identify your company's needs and the resulting actions that must be taken by the suppliers. Schedules can be discussed and agreed to. Workarounds, where necessary (and there will be a lot of them) can be formulated and agreed upon.

These vendor meetings also provide an excellent forum for your purchasing people to meet their suppliers face to face. The ensuing relations are usually enhanced when people know each other personally. Occasionally, extraordinary efforts are put forth by vendors and suppliers when needed out of respect for and commitment to your 'people', whereas they may not feel the urge to respond in a similar fashion for your 'company'.

Another benefit provided by the transition team is the regular opportunity for the team to discuss current engineering change activity and to plan for the timely cut-in of approved changes during the transition stage. Of course, you must have documented procedures for the manufacturing cut-in of approved changes during the production phase of your programme. Change cut-in during production should be straightforward and relatively easy to execute. However, the process of cutting in changes during the transition to production is often a different matter.

Agreements should be made with your customer, as discussed in Chapter 2, to permit your internal CCB to control changes to 'issued' drawings/designs at this point in the programme. The customer should not get involved in the act until after the product baseline has been established, i.e., until after the FCA/PCA.

With this in mind, procedures can be put in place to cut-in CCB (internal) approved design changes with appropriate controls and improved efficiencies. *Note*: Our PDM will serve us in good stead in this area.

Customer involvement

With few exceptions, your customers should not get involved in your transition activities until the completion of the FCA/PCA, i.e., the establishment of the product baseline. At this point, you will present your customer with the drawing package or equivalent for authentication, if a military product is involved. The authentication process is described below. If your product is commercial, you will simply continue on with production activities and maintain your drawing package in accordance with your Configuration Management Plan.

One of the exceptions to the absence of customer involvement is the 'initialling' of the engineering drawings prior to their 'issue'. This activity has two purposes:

1 To involve your customer in your development process for their own feeling of participation, i.e., the feeling of being 'part of the process'.
2 To allow them to feel that they are performing a 'sanity check' on your process before you go off and order production materials which may represent a significant portion of the total programme cost.

Admittedly, these intentional allowances of customer intervention, whether forced (by the customer) or voluntary, take time and may or may not insert added value into the process, they certainly enhance the feeling of good will between both your company's personnel and your customer. This good will is priceless. You will have made an important investment in the future of your programme.

One area that customers almost always participate in is the final design review. In Figure 5.1, design review 4 represents this activity. This is the culmination of the verification of your development process. Your FCA has been completed, and the test results demonstrate that you have met all of your system level requirements. Your customer will want to participate in this joyous (or maybe not) activity.

That's OK – either way. You must always lay your cards on the table for your customer to examine and, in many cases, chew up, along with some of your personnel, if the circumstances warrant it. That is part of the process. Get used to it because if your designs and test results can't stand the light of day (and your customer's piercing eyes), then you should find a gentler business with which to bring home the bacon.

Generally, though, your customer will support you and your employees in every possible way. They too want your programme to succeed. They have as large a stake in your success as do you and your employees. It's surprising how often your customer can help as the programme matures and cycles through its production, deployment, maintenance, and support phases.

Lessons learned

One of the benefits of gathering experienced, 'hands-on', people in a room to discuss the transition to production is that each brings their own unique 'lessons learned' to the meeting.

The test engineer has recently learned much about the goodness of the design from conducting the current or previous design evaluation and qualification tests. The manufacturing representatives know what went right and what went wrong during previous transition periods. As successive programmes evolve, mature and decline, additional insight is gained into the pluses and minuses, the good and the bad, the do's and the don'ts, to be applied to the next transition activity. It's a good idea to document these 'lessons learned' for future programmes and personnel.

FCA/PCA

I think that I have adequately described what the FCA and PCA activities are and have provided acceptable guidelines as to best practices for their conduct. The point I shall try to make in this chapter is that it is important to clarify and reaffirm the roles and responsibilities of the individuals and functional organizations that conduct and support these CM activities.

Once these roles and responsibilities have been 'signed up to' by those responsible, the specific actions and processes can be itemized and incorporated into the transition team's action item list. From there on, it's a piece of cake.

As I have said before, the FCA is conducted on the engineering prototype. The PCA is conducted on the first production unit. Remember that the purpose of the FCA is to demonstrate to yourselves and to your customer that your design works, i.e., that it meets your system level specification performance requirements. The purpose of the PCA is to verify that your manufacturing planning, which was created from the engineering design data and engineering drawings, produced a unit whose physical characteristics match the drawings.

One additional piece of advice on FCA and PCA. Get your customer to sign up to your FCA plan and your PCA plan as early in the transition phase of the programme as you can. This will allow both your customer and yourselves to negotiate the 'right amount' of FCA and PCA. Overkills in this area can be costly and underachievement can allow important information to drop into the crack, causing additional expense down the line.

Oh, and don't forget to conduct an Acceptance Test on your PCA unit to prove that it is the same physical *and* functional configuration as your engineering prototype. After all, you could prove that an artillery piece fires properly (FCA) and that you can build a nice toaster (PCA). I wouldn't want to go into battle with your toaster as my weapon, though.

The Acceptance Test, a subset of your design evaluation and qualification tests, will preclude this unfortunate turn of events from occurring.

Authentication

Authentication is the formal acceptance of the design by a military customer. In earlier times, the customer would wait until all aspects of the development activities were completed and then review the drawings for authentication. As you have learned in this chapter and in earlier chapters, because of shortened development and production cycles, the customer may have become involved at the time of issue of your designs. However, it would not be fair for us to request that they authenticate your design disclosure documentation before your FCA/PCA is completed and your product baseline is established.

Therefore, this is the way it is done! We complete the FCA and PCA, and then we request authentication. From that point forward, the customer participates in the review of all proposed changes. *Note*: On some contracts, the customer may request to be part of the review process for Class 1 changes only.

Schedule

Last but not least, is the schedule for your transition to production activities and milestones. Schedules are important in all programme phases but so many diverse activities occur during this phase that it is absolutely critical to generate, maintain and be accountable to the transition schedule.

Put your project engineer in charge of your schedule.

The last event in the transition to production schedule is the shipment of the first production unit to your customer. If you did your CM job right, it will be a good one, and they will be happy. That's what we all want.

6

Production and support

CM during production is different from CM during development and CM during the transition to production. As we move closer to our goal of excellence in CM, and integrate our PDM systems into our businesses, we will reduce these differences. After all, the basic idea is to capture and control information and to make that information readily available to those who want and need to use it.

CM during production should represent an institutionalization of stable processes. Problems are identified and resolved. One type of resolution is a design change. The design change approval 'vehicle' (in the old world – paper, in our new world – data displayed on computer screens) is circulated for review and approval or disapproval. The designs (or drawings) are updated, and the change is cut into production and (maybe) back fitted into previously built and delivered units.

In this chapter, I will describe these 'stable processes'. We will focus on what CM processes and activities are utilized during the production and support phases of the programme and their interrelationships.

Several of the topics that I will touch on in this chapter are explained in greater detail in other chapters. My purpose in mentioning them here is to bring them into focus in the phase of the programme in which they are utilized. In other words, these are production or support activities. In this chapter, I will 'pull together' these CM and related activities and provide pointers to where they are described in detail.

I think that this is an appropriate place to discuss the fact that most books on CM explain 'what' the various CM activities are and group these activities into four standard categories, i.e., Configuration Identification, Configuration Control, Configuration Status Accounting, and Configuration Audits. This categorization is not necessarily the order in which these CM activities are conducted during the course of a programme or project. Although I have adhered to this categorization throughout Chapter 4 I have endeavoured to present topics to you in the most sensible and understandable order to facilitate the realization of our goal of excellence in CM for your businesses.

CMP

Your CMP is the backbone of the CM process during the production and support phases of your programmes. If you have generated a practical CMP (no pun intended), based upon the checklists provided in Chapter 10 and the guidelines presented in Appendix A, and secured internal functional ownership for the processes and procedures specified therein, and if you have obtained buy-in and approvals from your customer, you have established the basis for a successful CM programme.

Remember the CM Planning Schedule we discussed in Chapter 2? Well, by now you have utilized it to capture your design baselines and to conduct your FCA and PCA. Your product baseline has been established. Your internal CCB members have been selected, and your CCB has already had several meetings to deal with proposed changes to your system level specifications and detailed designs.

Now, the procedure which you have identified in the Configuration Control section of your CMP will guide you through the change process for future proposed design changes. The CSA reporting activities in the Status Accounting section will provide you visibility into, and allow you to report on, the progress of those changes, and the Verification Audits described in the Audits section will determine if your Major (Class 1) changes have been cut into your production line at the planned times or on the specific units identified by your internal CCB.

Configuration Control Boards (CCB)

As stated in the previous paragraphs, you will have established your internal and external CCBs. Your internal CCBs (hardware and software) will review and approve or disapprove proposed changes to your baselined designs, and the results will be forwarded to your external CCB (if applicable, depending upon the scenario you negotiated with your customer at the beginning of your programme and documented in your CMP) or to your customer's representative for their approval. Figure 6.1 shows the recommended CCB reporting hierarchy.

Once your new PDM system is in place, and you have established on-line CCB functionality, the turn-around time for the review and approval process will decrease dramatically.

Major (Class 1) changes

The old Military Standards (MIL STD 480, 483, and 973) allocated many pages and even entire chapters to explanations of Class 1 and Class 2 changes. These classifications have recently been renamed to 'Major' and 'Minor'

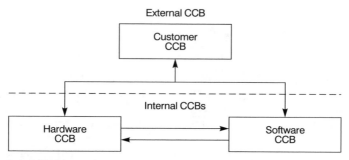

Figure 6.1 CCB hierarchy

design changes. I will try to cover that same ground in two paragraphs. This is pretty standard stuff, and if you want more detailed definitions, just pick up a copy of MIL STD 973 and read the section on Change Control.

I do not intend to minimize the importance of these change classifications. They are among the most important elements in CM. I do, however, want to keep things simple in this book. After all, if it's worth doing or worth understanding, it ought to be able to be explained in simple terms and in a few sentences. The truth is always simple (Lyon's Law).

A Major (Class 1) change is a design change that is absolutely required for the item in question to function as defined by its requirement specification. All previously built parts must be removed from service and either upgraded to the new design or scrapped, i.e., non-interchangeability is unacceptable. A Major (Class 1) change could also be one that affects cost and/or schedule. The production cut-in of Major (Class 1) changes shall be verified by adding the affected part(s) to the list of items to be verified during your Configuration Verification audits.

Minor (Class 2) changes are all changes that are not Major (Class1) changes.

Material Review Board (MRB)

During the course of receiving, fabricating, assembling, testing, delivering and installing your products, the occasion will occur (not often, we hope) where you encounter materials and/or parts and assemblies that do not conform to your engineering drawings and specifications. This non-conforming material must be dispositioned according to the rules and guidelines specified in your CMP.

The categories of non-conforming material are:

1 Repair (to drawing) in accordance with a standard repair procedure;
2 Return to vendor for repair (to drawing) or replacement;

3 Scrap;
4 Rework to an acceptable condition for specific use;
5 Use as is.

The first three categories may be dispositioned by a Material Review Team (MRT), i.e., a team of your company's employees who are allocated the responsibility for making the decision as to the appropriate disposition for your non-conforming material, and, if your CMP so specifies, a customer representative.

However, if after evaluation of the non-conforming material, it appears that Category 4 or 5 is appropriate, then review and approval or disapproval by a Material Review Board (MRB) is mandatory.

The MRB should consist of representatives from engineering, manufacturing, quality assurance, producibility, sourcing, and configuration management, plus customer representatives as (or if) specified in your CMP.

The function of the MRB is not limited to review and approval of recommended categorization but also to determine corrective action to preclude this type of non-conformance from reoccurring in the future. This corrective action is currently documented on Request for Waiver (RFW) and Request for Deviation (RFD) forms for military programmes as shown in MIL STD 973. As you implement the functionality of your PDM system, the data that are captured on these forms will be inputted instead, to computer screens and managed along with your design, contractual and procedural data by your PDM.

Suffice it to say, in concluding our discussion on non-conforming material, that once disposition of the material has been approved and appropriate correction recommended and approved by your MRB, then a system for follow-up to assure proper implementation of these corrective actions is needed. Of course, in the future, your PDM will provide the visibility and control necessary to satisfy this requirement. For now, though, you must rely on the intervention of your programme and/or project engineers, supported by your quality assurance staff to assure proper closeout of corrective actions. By the way, one possible corrective action is the initiation of a design change. This would trigger an input to your CCB.

Problem Resolution Board (PRB)

Chapter 8 will deal in length with the identification and resolution of problems that occur during all phases of your programmes. The point that I need to make in this chapter on production and support is that, after you establish your product baseline, there must be a system in place to identify and resolve problems that occur.

These problems may be identified by designers, builders, users, and maintainers of your products. They may also be identified by your customers.

They may occur at any time and at any place. They must be dealt with in a timely manner to minimize risks and costs to your programmes.

Configuration Verification Audits (CVA)

CVAs are usually conducted on each production unit supplied by your company if the product is a military procurement. If not, the option of whether or not to bear the cost involved with the capture and processing of CVA data is a decision which your company must make.

You must trade-off the cost versus the value of obtaining traceability data. Chapter 4 provided the categories and types of information obtained during these audits and provided the rationale behind this activity.

Remember during your deliberations that one of the benefits derived from conducting CVAs is that you will have verification as to whether or not your Major (Class 1) changes have been properly cut into production.

Automated CM

The detailed discussion of Automated CM is presented in Chapter 9. However, it is worth noting here that many of the CM activities which we have identified as occurring during the production and support phases of our programmes will be more efficient, some by orders of magnitude, when you have implemented your PDM system into your business operations and environment.

Support

Thus far, I have been directing your attention primarily to the CM activities conducted during the production phase of the programme. The same CM activities will continue throughout the support phase of your programmes, with additional emphasis on the problem identification and resolution activity.

Actually, there is such a significant overlap between the production and support phases of a programme that you will barely notice a change in the way in which you are conducting your CM activities.

7
Software and firmware

My approach to the Configuration Management of software and firmware will be to compare the processes utilized to develop and produce these products to hardware CM processes and procedures and then to identify similarities and differences. As you have seen thus far, my purpose is to identify and define CM activities and to provide you with the means and knowledge to institutionalize a practical CM system in your business. At the same time I am providing you with a practical methodology for effectively and economically incorporating CM into your development and production environments and transitioning your CM system from paper to electronic control.

Similarities

There are more similarities than differences between hardware and software CM. Every process we have examined for hardware applies equally well to software. You will capture software baselines, control changes to those baselines, report on the status of proposed and approved changes, and conduct FCAs and PCAs. The implementation of your electronic PDM, although during a later phase in your PDM implementation programme, will lead to similar funtionalities. The computer 'screens' may record and display slightly different 'attributes' or 'metadata', i.e., information, but the processes and interrelationships, including the CM interaction with your development methodologies, will be the same.

In software CM, as in hardware CM, we need to think in terms of both the design, which we will capture and control, and the software code, which we will identify, i.e., mark with a Computer Software Configuration Item (CSCI) number, and carefully incorporate software code changes into. The software code is analogous to our product hardware.

Differences

There is one significant difference between hardware and software – the 'build' process. The software build process consists of assembling and linking all those little elements and units of code and 'building' or assembling a software program out of this bag of 'stuff'. It also consists of keeping track of how this all comes together, i.e., what is where? The details of what files are encapsulated and how your PDM is integrated with your software build tools may vary but this phenomenon is true also when integrating different hardware design tools (especially the mechanical model tools) with your PDM system.

Most software development operations already have in place a process to accomplish some degree of data and code management. The question is: shall we, as CM people, support this function during development (Work in Progress) with our wonderful new PDM, or shall we get into the act only after the transition to formal internal change control? This is a judgement call that can be best resolved only with careful consideration and close communication between your CM and software development people. My advice is to start slow, i.e., capture and control only the finished product at first (validated software program) and then to move on to supporting the WIP process.

The software build process is, of course, analogous to the hardware 'build' or production process. The main difference is that the hardware build process occurs over and over again, whereas the software build process happens only once (if we get it right the first time).

I suggest that you first integrate parts of your software development process, e.g., scan in your software problem reports and use your PDM view tool to view them and print them out (I can't believe I just used the word 'print' in this book). Next, scan in your design review and development folders.

Next, capture your programming files. You may have to encapsulate these files before checking them into your PDM. Finally, start the process of the 'tight integration' of your software design tools with your PDM. You may have to wait until an appropriate interface module is available on the market before you start this activity. On the other hand, you may want to have a go at designing your own customization.

Firmware

Firmware is software which resides on a piece of hardware, simple as that! The software code is provided with a CSCI number, and it is then installed in a hardware device called a programmable read only memory (PROM) or an erasable PROM (E-PROM) or any of several similar type devices.

The basic idea is the same for all firmware. Treat the PROM as a piece of hardware and point to the CSCI for identification of its contents, i.e., the software program. Describe this process in your Hardware and Software Configuration Management Plans. Include in these documents the procedures you utilize to 'burn' the PROMS and to incorporate updates as changes to the firmware are approved by your CCB.

A more challenging question is the methodology to be employed to control dynamic changes to firmware which has been sold off and deployed in equipment in the field. New technology is making this possible and, from a design re-use point of view, it may well be the way of the future. The idea is to employ one hardware design to serve as the basis for many firmware designs by redesigning and redeploying the software element of the firmware package in the field.

As long as you keep in mind the basic tenets of baseline capture and documentation and all of the elements of configuration control and status accounting, you should not have a problem with this scenario.

Communication

An easily overlooked CM activity is communication between the hardware and software CCBs. Items of shared interest often get brought up before one board and dispositioned there but if the other board is not notified or even questioned as to possible impact, then all manner of things may go wrong.

For example, if an electrical module is modified or replaced by the hardware CCB in order to correct design deficiencies or to improve performance, and the software in a PROM is not updated accordingly, then the module will fail acceptance testing or, worse, make its way out to the customer before the problem is identified and corrected. Down line costs to fix problems are always more expensive than doing it right the first time.

Communication problems can be avoided in two ways:

1 Have a software CCB representative sit on your hardware CCB and vice versa. This is the most sure-fire way to preclude one board from going off half-cocked. The software representative, by actually being present in the room when the hardware design change is being discussed, can determine for themselves whether or not there may be a software impact. The downside of this solution is the cost. We have just arranged for your software representative to spend approximately one hour per week in an additional meeting.
2 A second, and less expensive, way to provide adequate communication between the two internal CCBs is to assure the proper recording of minutes and action items. This report of activities and proposed actions, if provided to the other CCB in a timely manner, can allow for appropriate review and determination of impacts.

Another area of communication that can provide significant long-term benefits as well as preclude short-term problems is the sharing of 'lessons learned'.

As each CM activity, either hardware or software, solves a problem or develops a new process, this information, if communicated to the other CM activity, may very possibly also improve its efficiency. Sometimes, we get so caught up in the business pressures which affect us directly that we don't think of the other fellow. Well, here is one area where that kind of selfish attitude can seriously harm a business's performance and costs. Talk to each other. It's as simple as that.

Co-location is another way to improve communication. If you are sitting in the cubicle next to mine, chances are pretty good that I may pick up on something you are doing that may be applicable to my function as well. You may mention it in passing or I may just inadvertently overhear a conversation (not eavesdropping, of course – we never do that).

There is one more activity where we should assure good communication between the hardware and software elements of CM. That is during the CM planning phase of each programme or contract. Chapter 3 outlined the process for planning a CM programme. That process applies equally to software and hardware CM activities.

The scheduling of design reviews is usually carefully coordinated between the engineering units performing the hardware (mechanical and electrical) and software design activities. These design reviews are the events which we, as CM people, will use to transition to the next higher level of control. Therefore, we all need to talk to each other starting at the contract proposal kick-off and continuing throughout the programme in order to obtain ownership on the part of each other as to who will do what and when. We also do not want (ever) to assume anything. For example, if I assume you are going to have a design completed at a certain time, and I then assume I am going to put that design under CM control at that time, I am setting both myself and you up for major disappointment (and possibly worse).

Things just don't work out that way in business. Never did, and probably never will. That is the reason why the use of design reviews as transition events are so effective and reliable. Since a certain 'goodness' in the design can be expected and, in fact, demanded at each design review, we can depend upon these events to define outputs from engineering and inputs to CM.

Roles and responsibilities can be defined at, and for, each design review. That is why we must communicate often and well – CM to engineering and CM (hardware) to CM (software).

Organizational structure

A typical hardware and software CM organization is depicted in Figure 7.1. The hardware/software subdivision, if one exists, is usually at the group or team level. Sometimes these groups or teams report directly to a unit or function manager.

Sometimes the unit manager in charge of the CM function also has the drafting organization and/or the document control centre reporting to them. The placement of the drafting organization in an engineering operation varies. This scenario shows only one of the possible places into which to 'drop' the drafting organization. It is not within the scope of this book to make a

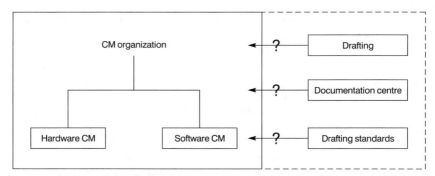

Figure 7.1 Hardware/software CM organization structure

recommendation in this area. It is, however, necessary to mention it because the definition of procedures as to who will deposit and withdraw files to and from your PDM staging areas involves decisions as to roles and responsibilities for drafting versus engineering personnel. As CAD design tools improve and as your PDM system becomes firmly established as the means to invoke your design tools and becomes the manager of your design data, the roles and responsibilities of engineers and draftspersons will change.

8

Problems and resolutions

After you establish your product baseline, you must have a system in place to identify and resolve problems that occur during the production and support phases of your programmes.

These problems may be identified by designers, builders, users, and maintainers of your products. They may also be identified by your customers and suppliers. They may occur at any time and at any place. These problems must be dealt with in a timely manner to minimize risks and costs to your programmes.

Figure 8.1 shows, at a high level, how a typical Problem Resolution Board (PRB) is integrated into a well organized engineering operation. Note that there are no limits to the sources of the potential problems. Note also the feedback mechanism put in place for verification of corrective actions.

As you can see, a subset of the possible corrective actions is a recommended design change. This recommendation is passed on to the appropriate CCB (hardware or software) for disposition. It is not within the scope of the PRB's responsibilities to approve design changes, only to recommend them to the CCB.

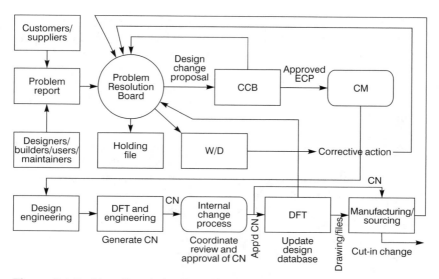

Figure 8.1 Problem Resolution Board/change flow process

Problem Resolution Board (PRB)

A PRB should be established to identify and resolve problems which arise during the production and support phases of your programmes. This board will take over the problem resolution responsibilities of your transition team as you move into full-scale production.

Mission

The mission of your PRB will be to receive, evaluate, investigate, and dispose of problems that are identified to it. Your PRB will recommend design change activity where appropriate. If it can, it will solve problems directly. It will document problem inputs and provide status and traceability information throughout the problem resolution process. It will report on the status of the problem resolution process to your management and to your customer as provided for in your company's internal procedures and as documented in your CMP.

Your PRB shall assure timely and effective closeout of all problems identified to it. It shall be responsible for the determination and consideration of technical, cost and schedule impacts to your programmes.

Membership

Your PRB should be composed of the following membership:

1 *Chairperson* – The chairperson shall preside over the PRB meetings. They shall assure that the PRB process is conducted as specified in your company's CMP and assure resolution of the issues brought before the PRB.
2 *Administrator* – The administrator shall prepare and distribute the PRB meeting agenda (including a list of current action items and responsible presenters), record action items, generate and publish PRB meeting minutes, and maintain the PRB problem database. The administrator shall also assign an identification number for each reported problem.
3 *Project engineer* – The project engineer shall be responsible for the evaluation of inputs to the PRB and the subsequent resolution and disposition of these problems from a technical perspective. They shall initiate design investigations as appropriate and shall monitor cost and schedule impacts.
4 *Functional representatives* – Representatives from the various functional organizations shall discuss the status of items they have been assigned by the PRB. They shall report on each topic under consideration by the PRB at the PRB meetings.

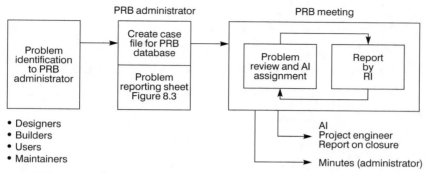

Figure 8.2 PRB process flow chart

Process

The PRB process is shown in Figure 8.2.

1 Problems shall be identified to the PCB administrator by the originator, i.e., designers, builders, users, or maintainers of the product or equipment. The problem shall be documented on a Problem Reporting Sheet (see Figure 8.3).

2 The PRB administrator shall assign a number to the problem and initiate a case file. They shall enter the appropriate data (excluding failure information) into the PRB database.

3 The administrator shall present the problem to the PRB at the meeting. Details shall be provided to the PRB by the individual(s) who identified the problem to the administrator.

4 The PRB will evaluate the problem to determine that a problem actually exists and, if so, determine the appropriate steps to take to resolve the problem. The PRB shall also categorize the problem for future metrics evaluations.

5 The PRB shall either solve the problem directly at the initial meeting or assign actions to individuals to drive towards a solution of the problem.

6 The individual assigned responsibility for an action shall report on the status of their activities at each subsequent meeting of the PRB until the problem is resolved or until their part in the solution is completed.

7 When all actions have been completed, and the problem is resolved, the administrator shall document the findings of the PRB in the database and shall issue the results in the meeting minutes.

8 The project engineer shall follow up on all actions directed by the PRB. If a design change is recommended by the PRB, the project engineer shall communicate the necessary information to the appropriate CCB.

9 The administrator shall record all assigned action items in the PRB database in a 'rolling' action item file, which they shall provide to the

PRB # _____ HW/SW/FW _____

Ref. Doc. # _____ Equipment _____

Problem identification source _____ Component _____

Presenter _____ Category _____

Problem _____

_____

_____

_____

Status/action _____

_____

_____

_____

_____

_____

_____

Close out action _____ Close out date _____

Close out approval _____

Figure 8.3 Problem Reporting Sheet

project engineer as a tool to utilize in the follow up on actions taken and closure status.

10 The project engineer shall provide feedback to the PRB regarding closure to all dispositions or recommendations issued by the PRB.

The timely identification, evaluation and resolution of problems during the production and support phase of your programmes will reduce costs and improve delivery schedules. Product performance and reliability will be enhanced, and you will maintain a good relationship with your customer.

Considering the problem resolution and disposition activities as the 'front end' to your change control CM process will help you to cover all bases and assure that nothing 'drops in the crack'.

9

Automated CM

Introduction

In this chapter, I will present the steps necessary to achieve our goal of automated CM. I will provide you with the information you need to plan and implement the functionality of a PDM system within your business. When you have conducted all the activities and operations described herein, you will have completed your journey to 'excellence in CM'. There will be additional improvements that can be made, and you will find updates to and expansions upon the information provided herein in Appendices B, C and D, but you will have the basics of automated CM in place. You can 'tweak' your CM/DM/ PDM system in the future as you deem appropriate.

That's quite a mouthful? Well, it's true. Wait and see. You have nothing to lose and everything to gain. It's a win–win situation. Here we go!

Figure 9.1 provides an overview of the PDM system that we are going to implement.

Note: Appendices B, C and D supplement the information contained in this chapter.

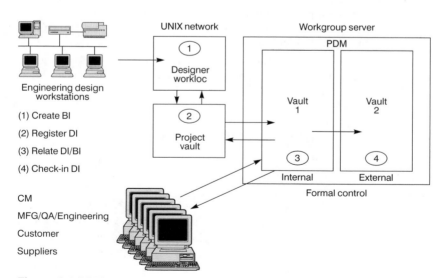

Figure 9.1 PDM overview

I suggest that you first read this chapter, then read the three appendices specified above and then re-read this chapter. There's a lot of information contained in these sections, and I don't think it is possible to absorb it all during the first reading. I have tried to organize this information logically between Chapter 9 text (overall, generic approach) and Appendices (additional specific, detailed information).

Planning

As in the establishment of our basic CM methodology and processes, it will be necessary to do quite a bit of up-front planning for our PDM system functionality implementation and process integration. We will accomplish our PDM implementation by increasing the functionality of our PDM system via a phased approach. Then, at the right time, we will integrate this functionality, in an incremental manner, into our business programmes. We will start our planning activity by first performing a requirements analysis, then by defining our desired operating environment and finally by documenting these requirements in a system level specification. Next, we will examine how we are going to handle legacy data and how we are going to integrate our PDM system with our current and future CAD and office tools. We will plan for the requisite documentation and training plus plan for the necessary hardware and software resources for our internal operations, our customers and our suppliers. Our planning activity will conclude with the generation of a detailed program schedule.

I recommend that you establish an appropriate working space for your PDM project team to conduct meetings, storyboard CM-related business processes, meet with potential PDM tool vendors, evaluate PDM and third-party tool software, troubleshoot problems, develop workflows, implement processes and procedures, discuss issues and concerns and report on progress. I recommend a 20×40 ft room, equipped with corkboard walls, vu-graph screen, PCs and workstations, telephones, tables and chairs. Call it your 'PDM Operations Centre' or some such thing.

PDM System Requirements Analysis

The first step in the planning process is the definition of your automated PDM system requirements, i.e., you must conduct a requirements analysis. The requirements analysis is simply the thinking out and documenting of the processes you use to conduct CM today and updating them for how you want to do CM in the future. This initial part should be easy – it's the 'best CM practices' defined in the first eight chapters of this book applied to your current CM-related business processes. You may say, 'well, I'm not doing CM that way, yet!' Of course, you are correct but I believe that the CM methodology proposed herein is the way you will want to go in the future, aided by your PDM system. So, let's start off by assuming you are mentally focused on the 'best CM practices' and are now about to automate them, OK?

You should start by conducting 'storyboard' sessions for all of your CM-related processes. Call in the actual 'doers' for specific tasks. Involve management and functional 'lead' personnel as appropriate. Have the 'doers' write down each step in a process on a card and pin the card on a corkboard in your PDM Operations Centre. Don't criticize any input. Just pin up all of the cards with actions written on them. You can reorganize and delete individual inputs later, as you discuss among yourselves the order of steps in any process.

Identify and examine each step of each process in excruciating detail. The time you spend doing this up front will preclude many 'repairs' to your PDM workflows down the line. This is the time to thoroughly understand your current processes and to fix problems as they are identified. You have the opportunity here to 'tweak' your processes so that you establish the 'best CM practices' for your business. Please remember that your new PDM system will be no better than the basic processes which you have/will put in place and which you are planning to automate as you migrate your business from paper to electronic control. Make sure that you obtain 'buy-in' from all participants in this process. Document the results of your storyboard sessions on flow charts.

Remember, you are capturing on paper the way you do business. Your PDM tool will eventually reach into all facets of your business. Even though you will be implementing PDM functionality a step at a time, you don't want to do anything today that will preclude implementing future functionality in other business areas. You will see what I mean as we proceed.

Once you have generated the flow charts defining your CM processes, the next step is to define your PDM data model 'objects', 'attributes' and relationships. These data model elements will be used to define input, query and result (report) screens. You will then generate 'workflows' for your PDM system. You will create user groups, define security access rules, create interrelationships, and develop rules, roles and vaulting scenarios. Remember that your PDM system will be utilized to manage the actual physical files to be captured and controlled and even manage your legacy paper, if you don't want to pay for the conversion of your legacy data into electronic files for induction into your PDM. Your attribute objects will define the data elements to be identified and managed. Your workflows will move files from personal work locations to vaults and vice versa and will promote files from one vault to another. They will also invoke different permissions and access rules as your data are promoted from state to state. They will invoke other software programs to convert file formats, i.e., from native formats (design tool native CAD files and office tool native files) to neutral formats (raster, IGES, PDF) for view capability and CALS data deliveries. Your workflows will move change notices and design markups from one member of your CCB to the next

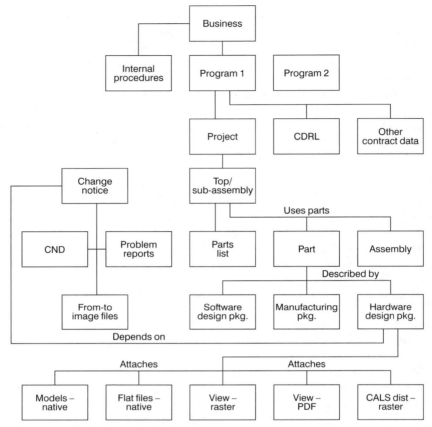

Figure 9.2 Data model – overview

and provide notification per your internal e-mail system. Ultimately, your PDM will be utilized to launch and control your design and office tools. Figure 9.2 shows a typical data model 'object' chart. Figure 9.3 provides an example of a set of attributes for a typical design package. These data represent the information which, in the old world, would have been present in a drawing's title block plus the data used by a documentation distribution activity to process and distribute drawings and parts lists.

Note: You will start your automated CM process by simply checking design files into and out of your CM library vaults as changes are approved by your CCBs. You will then implement on-line CCBs, controlled by your PDM workflows. Figure 9.4 presents a workflow for a typical on-line CCB. Finally, you will have your design engineer and draftspersons interface directly with their design tools through your PDM to effect changes directly upon receipt of approval from your on-line CCBs or prior to submittal of proposed changes to

Design document			
Attribute	*Type*	*Field*	*Description*
DD_No (key)	Character	25	Doc number (icon display)
DD_Rev (key)	Character	3	Revision
DD_Project	Character	30	Project name
DD_Cage_Org	Character	5	Doc CAGE
DD_Cage	Character	5	Design COG CAGE
DD_Title	Character	30	Doc Title
DD_Type	Character	10	Type
DD_Size	Character (value set)	1	Size
DD_Shts	Integer	3	Number of sheets
DD_Diststmt	Character	1	Distribution statement
DD_Distlstcd	Character	10	Distribution list code
DD_Rep	Character	25	Replace drawing number
DD_Contract	Character	30	Contract number
DD_Date	Date	8	Date
DD_Security	Character	30	Govt classification level
DD_Rights_Gt	Character	10	Govt rights
DD_Rights_Co	Character	30	Company classification level

Figure 9.3 Data model object attributes – hardware design package

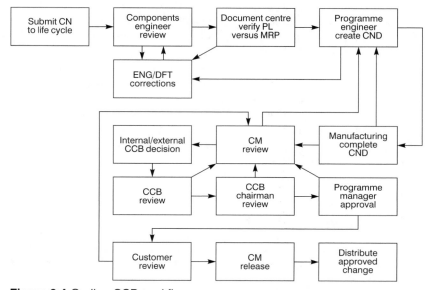

Figure 9.4 On-line CCB workflow

your on-line CCB workflow by creating the 'to' version of your design and including it in your change proposal package. There will be more on this subject later in this chapter. For now, I just want to introduce the concepts with which we will be dealing.

PDM System Requirements Specification

You now need to document the results of your requirements analysis in a PDM System Requirements Specification. This document will not only capture your PDM requirements but will also guide you through the implementation of your PDM. It will also allow you to test your PDM system to verify that you have met your initial system level requirements (sort of like a good old FCA?).

PDM tool selection

There are many PDMs on the market. I dare say that this number will increase exponentially in the near future. I would be willing to take bets on that.

There are two main classifications of PDMs – 'Commercial-off-the-Shelf (COTS)' and 'toolkits'. The COTS variety, though initially easier to implement, may be more limited in their functionality and flexibility than the toolkits. Some COTS products are pretty good, though. Which type of PDM tool you choose will depend upon your business requirements. I feel compelled to advise, though, that the better PDM toolkits permit the implementation of greater functionality, integration into virtually limitless variations of business scenarios plus provide a more robust 'engine', i.e., greater functionality capability for modest investments in implementation efforts, integration set-up time and administration and maintenance level of effort. Refer to Appendix C, PDM tool and vendor evaluation and selection process, for a detailed treatment of this subject. You can find a listing of PDM and third-party tools and vendors plus lots of other good CM 'stuff' in the internet CM *Yellow Pages* at URL: http://www.cs.colorado.edu/users/andre/configuration_management.html

Implementation strategy

We will select a PDM implementation strategy which allows us to incrementally add PDM functionality in parallel with our ongoing CM activities so that we can continue to support our programmes and assure our customers that there will be no interruptions or delays. The essence of our strategy will be to install and implement the functionality of our PDM in four phases:

1 File check-in and check-out;
2 View, print, raster edit, distribution, backfile conversion, on-line CCB;
3 Hardware design/office tool integration;
4 Software design tool integration and CITIS.

We will also phase the use of our PDM into our various product lines, programmes and projects on a planned and agreed upon schedule in order to take advantage of our new automated capabilities in as timely a fashion as possible while not jeopardizing development or production activities.

Verification of functionality

Proceeding to integrate our new PDM into our programs without first verifying its functionality would be irresponsible and could be disastrous. So, we will check out each separate bit of functionality, each keystroke, each file transfer, each data entry, each user access attempt, each image file view and/or print operation, each file conversion process, etc., using a 'test environment' instead of the real 'production' environment before we start to capture and process real data.

Documentation

You must have plans and internal procedures for all of your business activities if you are going to conduct business with the DoD. If you are strictly a commercial house, then it is still a good idea to have written procedures to define your processes. You may think that you have the bases covered if you have highly motivated and competent employees but you will be in trouble if one or more of them gets in trouble, e.g., becomes ill or disabled for any number of reasons, and you need a backup. Accurate, simple, step-by-step procedures are indispensable in today's business environment where we don't have the luxury of the larger staff numbers permitted by the economies of the 1970s and 1980s.

Training

You should plan to conduct two levels of training for your employees and customers:

1 Overview training;
2 Detailed, hands-on, training.

The overview training is meant to spread the word that there is a new fellow on the block, i.e., your PDM. Remember, this is going to represent a culture

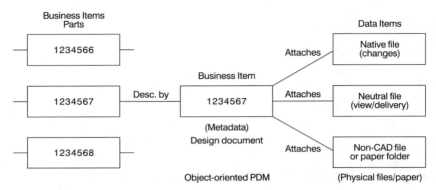

Figure 9.5 How it works

change (or shock) for many and a mind-set modification for others. Not everyone needs the same type or amount of detailed, hands-on training but everyone needs and will benefit from hearing about this new system that will, in many ways, affect their jobs and their duties. Refer to Figure 9.5 for an example of the type of vu-graph you should use in your training programme.

The detailed, hands-on, training should be administered in three categories:

1 CCB member training;
2 Designer/draftsperson training;
3 Casual user training.

The CCB member training should consist of approximately six hours of PDM theory, CM process review and hands-on PDM operation. CCB members must learn how to query for design and manufacturing data, route change packages around for review and approval, view, markup and review on-line images, and respond to messages.

Designers and draftspersons must learn how to launch their design tools via the PDM, check out design files for update and check them back in, and submit change packages to your PDM workflow processes. As we shall see later in this chapter, the ultimate goal is to enter data once, update it within the PDM/design tool system and use these data many times by internal and external functions plus your customers and suppliers. All designers and draftspersons must know how to at least enter and exit the system, manipulate files and query the system for information. Their training course should take approximately eight hours.

The casual user is the individual who only wants to view and/or print a drawing or document. They would previously have gone to the documentation distribution centre to request a print of the desired drawing or document. They will use the 'casual' view functionality of the PDM. Their training will only amount to two hours.

Resources

You will need the following hardware and software resources to meet the requirements of the more sophisticated PDM systems:

1 *Hardware* – There are three platforms for now and the immediate future that you should consider or must consider supporting, depending upon the computing resource mix in your own business:

(a) UNIX Workstations (SUN, HP, etc);
(b) IBM compatible Personal Computers (PC);
(c) Macintosh PCs.

Most robust PDMs require a pentium (or better) PC plus 32 MB RAM and approximately 100 MB disk space (for temporary storage of images). You will also need a high-speed connection to the computer network which supports your design tools.

2 *Software* – The primary software application required is the PDM tool, itself. Sufficient licences will be required to support your concurrent users and/or individual client workstations/PCs. I recommend procuring the minimum (estimated) number of seats to support each level of functionality as you proceed in your implementation, and later increasing the number of seats as your requirements grow. Third-party software may be required to support viewing drawings and documents, markup, from–to comparisons and edit capabilities, conversion of files from one format to another, e.g., native to raster or PDF for viewing documents, native to neutral file formats (raster, PDF, IGES) for CALS deliveries, on-line conferencing, etc. The extent to which you require third-party software will vary in proportion to the level of automation implemented and the degree of sophistication required of your PDM system. I will lead you through the implementation process for a typical, yet comprehensive PDM system in this chapter. We will discuss the hardware and software resources required for each element of PDM functionality as we proceed along our path to automated CM. One final note, while we are still in the planning mode – don't forget to plan for software maintenance, also.

Program integration

When we complete the implementation of Phase 1, file check-in/check-out functionality, we will be in a position to start capturing digital design baselines.

Chapter 4 described the baseline capture process. What we need to discuss at this point is the process of establishing the PDM functionality that you need to allow you to implement those processes. You will accomplish this by approaching the key individuals who conduct and control each of your programmes and projects in order to discuss and agree upon the extent of integration appropriate for each programme, both for legacy data and for current and future native CAD data. The trade-off of cost versus control is a key factor in this and most areas of PDM implementation and integration. We will examine the legacy data part of this subject in detail in the section on legacy data which follows later in this chapter.

The point here is to address all your programs and put your cards on the table. Make the 'who, what, and when' decisions and document significant milestones on a program schedule. It is not necessary, or even wise, to try to capture all program data immediately. Do it one step at a time, one program at time. The 'lessons learned' from each program will increase your progress up your learning curve at a surprising rate. Don't forget to discuss contractual data (deliverable and non-deliverable) and product user and maintenance documentation as well. Finally, decide whether or not you want to capture and control your business's internal operating procedures. Don't worry, though. If you don't do it now, you can do it later.

Schedule

One of the single most important tools you will need to support the establishment of your PDM system is an implementation and integration schedule. Start right away to create one by using the checklists in Chapter 10 to identify the major and minor tasks and activities required.

Don't worry about making mistakes and selecting too many or too few tasks. Just do it! I can't stress this enough. Your PDM implementation schedule will become a living document, and it will change as your proceed. My key message is to create a 'strawman' schedule as early as possible. See Figure 9.6 for a sample PDM implementation schedule. Put in estimated dates. I don't care how crazy they may seem to you at this point. You will have plenty of time to 'get it right'. The very act of creating your schedule will force you to think about the things you have to do. It will require you to talk to your people and to solicit their input. This activity in itself will generate invaluable information. Every hour you spend on your schedule will provide you with many happy rewards down the road.

Your PDM schedule, in conjunction with your action item list (see below) will keep you out of a good deal of trouble as you journey along your path towards excellence in CM.

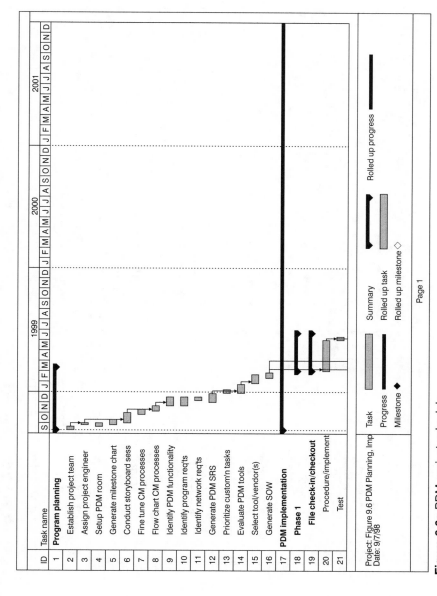

Figure 9.6a PDM project schedule

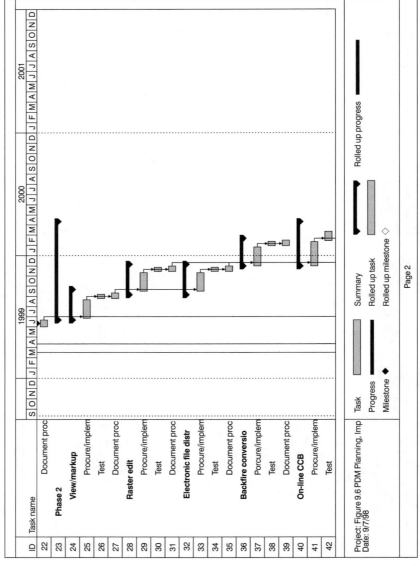

Figure 9.6b PDM project schedule

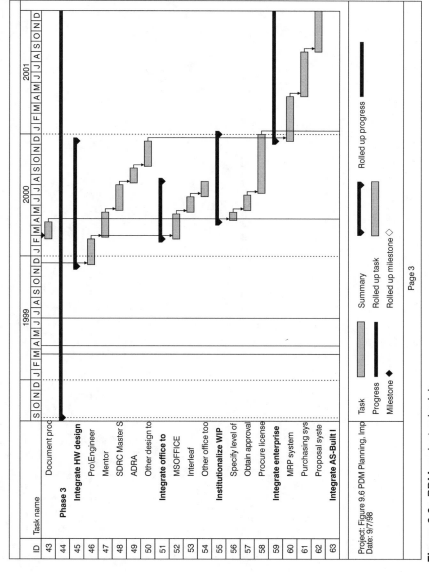

Figure 9.6c PDM project schedule

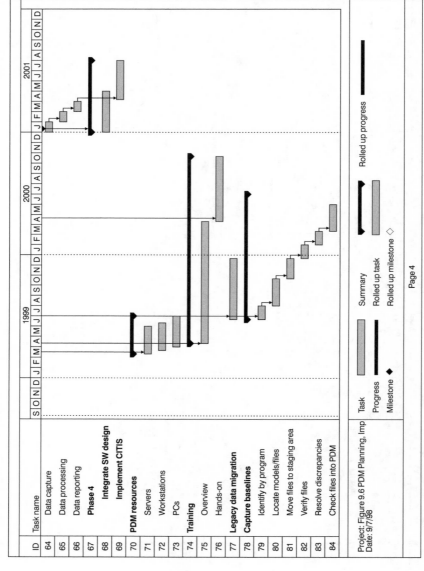

Figure 9.6d PDM project schedule

PDM project team

As soon as possible after your decision to proceed with your PDM implementation, you should establish a team of individuals to plan, coordinate, monitor, procure, implement, integrate, provide training for and document your PDM system. This team should, as in the case of your CCB, MRB, PRB, and transition teams, represent all involved and affected personnel. You may decide to institute a 'core' team of key individuals, with supporting personnel called upon to participate as appropriate. In addition to the users of your PDM, your PDM project team will have as members those personnel who set up and maintain your computer systems and networks, install and maintain your design tool and office computer software, and run your current CM operation and data distribution centre. Management must be closely involved, because your PDM system will represent a significant initial and follow-on investment, both in terms of material cost and manpower. They will want to see what they are getting for their money, and they should be active players in this venture. Above all, you will want and need their 'buy-in'. Their initial investment will be repaid many times over if your project team does its job right.

The PDM project team should meet weekly to discuss current status, resolve problems, make decisions, and follow up on action items. This team is the key to your success in establishing and institutionalizing your PDM system.

PDM project engineer

Your PDM project engineer will be the cornerstone of this new venture. They will be the hub around which the spokes of your PDM 'wheel' revolve, at least until its functionality is firmly established and all programs are successfully integrated and utilizing its functionality. Your PDM project engineer should be knowledgeable in CM and data management processes plus be familiar with the design tools, computer resources and networks utilized by your organization. They should ideally have experience in all functional areas within your business, and have a successful track record of completing projects on time, within programme budgets and within successful performance criteria. They must keep the project team fully informed as to the progress of the PDM installation process and its integration into the various programme areas. They should be the individual responsible for assuring that all action items are clearly articulated and assigned to specific individuals with a definite closure date identified. They must drive these action items to completion within the allotted time.

Above all, your project engineer should be a dedicated individual who demonstrates 'ownership' for all aspects of this new venture. You should be able to trust them to successfully lead you along your path to excellence in CM.

PDM project team members

The core members of your PDM project team should include, but not be limited to, representatives from the following functional areas:

- Design engineering – hardware and software (lead plus management personnel);
- Drafting;
- Computer resources (information systems), including network systems;
- Design tool installation, integration and maintenance personnel;
- Configuration and data management;
- Data distribution centre.

Supporting personnel from manufacturing and quality assurance operations plus field and product service, reliability engineering and documentation services should be called upon to support the project as required.

PDM implementation and integration

We will be moving along two paths simultaneously as we journey down our road to excellence in CM. First, we will implement some degree of PDM functionality. Then, we will integrate that functionality into one or two programmes. You will continue to implement functionality in an incremental fashion and then integrate this functionality into your programmes until full PDM functionality is achieved and all programmes are on board for legacy and current design data plus contract data and internal procedures.

We will install our PDM functionality in four phases as follows:

Phase 1:
- File check-in/check-out.

Phase 2:
- View/markup;
- Raster edit;
- Distribution (CALS);
- Backfile conversion;
- On-line CCB.

Phase 3:
- Integrate hardware design and office tools;
- Institutionalize WIP;
- Integrate MRP tools;
- Integrate as-built process.

Phase 4:
- Integrate software design tools;
- Implement CITIS.

Phase 1

Phase 1 is the jumping off place from our old system of trying to manage design data by capturing paper or mylar drawings, placing them in a controlled vault, and approving changes to them via our CCB. We are, of course, kidding ourselves today in thinking that, if we control the paper, we control the design. The 'real' design master resides in the design tool or database that created it. It is those native CAD files which our engineers and draftspersons will modify in the future to incorporate changes to our products. The paper, which merely serves as a medium to contain a plotted image of the design, is really only a copy. The digital design files are the true 'masters'.

File check-in/check-out

Figure 4.1 provided an illustration of the elements required for the check-in and check-out of design files. Chapter 4 provided the process definition. What we must discuss in this chapter are the steps you will have to take to put the hardware and software resources in place to implement this process.

First, select servers and networks to support this initial implementation. You should plan on a distributed network and database to allow for future expansion and to enhance system throughput. Make sure you have established TCP/IP addresses for these workstations and PCs connected to your network that you now want connected to your PDM system. A little trial and error is OK. Set up your PCs as 'clients' (as compared to running as an external terminal). This will save you grief in the future when you have all or most of your engineers running their design tools through your PDM, causing potential network traffic jams.

Most good PDMs support a variety and a mix of workstations, PCs and MACs. If you are using today's typical electrical and mechanical design tools you should have no interface problems, except for the capture and control of mechanical design models and some electrical design files. We will discuss this 'challenge' in greater detail later on in this chapter.

Install your PDM database and application software, using the installation procedures supplied with the product. Answer 'yes' when you are asked whether or not you wish to set up a distributed environment.

We will assume that you have chosen an object-oriented PDM toolkit (versus an 'off-the-shelf' product) in order to realize the full benefits of power and flexibility. *Note*: If you decide to purchase an 'off-the-shelf' product, you can still use the following functionality implementation procedures as a guideline, along with your tool's specific instruction set.

You will need to create the objects, attributes and relationships necessary to support your initial functionality, i.e., file check-in/check-out. Refer to Figures 9.2 and 9.3. *Note*: You will not need to generate any workflows at this point in time. The only unique object which you will have to create for file check-in/check-out is the 'hardware design package'. This object is a Business Item (BI) and contains only metadata (information). The attached Data Items (models – native CAD, flat files – native CAD, view – raster, view – PDF, and CALS distribution – raster) are objects that represent the actual physical files that you will be checking in and out and copying out of your PDM vault. Your PDM product documentation will provide procedures for the creation of these Data Item (DI) objects. You can use the 'out-of-the-box' functionality for your DI.

In order to create your hardware design package BI, you will have to decide what information you wish to capture concerning your designs. Think in terms of those data that you will want to perform queries on. Remember that soon you will be able to view drawings and documents. You can see any information that is contained therein so you don't have to enter it twice. Data entry costs money. Keep it to a minimum. You can always make adjustments later. You can use the sample attributes listed in Figure 9.3 as an example. Add or delete data elements as you desire.

Once you have created your BI and DI, you can try out your PDM system. It's a good idea to get used to the 'feel' of your system first, though. Perform a few queries. Run through the 'out-of-the-box' menus, actions, queries, file management procedures and read the help topics. Then you can create a personal work location and validate yourself as a user.

Move a file to your personal work location. Create a BI and a DI for it. Drag and drop the DI on the BI to establish a relationship between your BI and DI. Then check them into a PDM vault

Simple as that!

Continue to capture a small project's worth of design data. Create BI and DI for each design. Build a family tree for your parts and assemblies. When you have a baseline established in your PDM, utilize these files to update your designs as changes are approved by your CCBs using the procedures provided in Chapter 4.

Phase 2

Phase 2 of our PDM system implementation adds functionality which allows us to view designs (drawings) and other documents (contract data, internal procedures, specifications, problem reports, plans, etc.), route proposed changes around for approval (on-line CCB), convert native CAD design data to viewable and deliverable neutral formats (MIL STD 1840), and convert legacy data to formats acceptable for view, storage, distribution and management within our PDM system (backfile conversion).

When you have incorporated this functionality, you will be able to capture and control baselines exclusive of your obsolete paper system.

Some third-party software may be required for the view, markup and file conversion functionality. It is appropriate to mention this fact up front, since some degree of integration among software packages may be required, and this takes planning and, maybe, the services of an external system integration operation.

View

You will need to consider two categories of view functionality:

1 PDM view;
2 Casual view.

PDM view is required by designers, CCB members, and anyone else who will need to view and check-in/check-out drawings and documents contained in and controlled by your PDM system. You will achieve this capability either via the built-in functionality of your PDM tool or via an add-on third-party view tool. *Note*: If a third-party view tool is required, special customization may be necessary to interface your view tool with your PDM.

Casual viewers are those individuals who now walk over to your documentation distribution centres to order copies of drawings, parts lists, wiring lists and other documentation (or order them via computer terminal or telephone). All these personnel really need is to look at and, possibly, print out a copy of these documents. They will not be checking out design models or design data for update or modification or approving changes.

Your system will have to provide for the proper security checks in either case, however, to assure that only authorized personnel are allowed to view documents. This is another reason for the 'tree' structure of the data model shown in Figure 9.2. Each individual must initially be validated through a series of objects, and their validation data must be stored in your PDM system's database to be used in the future for re-validation prior to allowing them to view or print out documents, whether via your PDM view or casual view functionality.

Casual view is really a combination of a general purpose database (PC and/ or workstation) to allow for user identification (security access) and query for the desired document, plus a view tool to observe the desired image. *Note*: We will utilize another third-party tool to convert our design data that we will store in its native format for future changes to a neutral format for viewing and distribution. This neutral format will be CALS Type 1 raster, i.e., CCITTG4. We will soon probably be using Portable Data Format (PDF) files for viewing and distribution of textual data but we are not quite ready for that, yet. We will stick to raster images for the time being because of their reliability, acceptability and the availability of several view and print tools which accommodate the raster format.

Markup

You will need an annotation tool to highlight and markup the change package 'from–to' design images which you will send around for review as part of your on-line CCB process. The markup tools readily available on the market today include colour options to identify the various participants and can create markup 'overlay files' which can be stored as part of the permanent change package history.

Raster edit

You will require a raster edit tool to make changes to your legacy or current drawings or other documents which you have converted from paper, mylar and aperture cards to electronic files via raster scanning. Of course, native CAD files, captured for control of your current designs will be utilized for future modifications and updates. *Note*: I recommend that all legacy data be raster scanned rather than to try to capture old CAD files and the old design tool software versions that these designs were created on. The optional scenario of old CAD tool version management could present monumental problems when you consider the ramifications of possible non-backwards compatibility, future network change impacts and licensing impacts, e.g., 'will my old CAD data be compatible with my new design tool software version?', 'will it work with my new design tools on my new network configuration?' and 'am I licensed to use both my new and my old versions of software?'. Make life easy – raster scan your old drawings. Store these images in your new PDM system, and update them with a raster editing tool. There are several good ones around. If you want to also capture and control some (or all) of your legacy native CAD files, you can certainly do that and then decide which file (raster or native CAD) you want to update when the time comes for a design change.

Distribution

Distribution of electronic data to the DoD and its contractors and suppliers must currently exhibit compliance to MIL STD 1840. If you wish to do business with the DoD in the future, your data will have to comply with the requirements of MIL STD 2549. You have several options, here, but the simplest is to convert your files to raster (we have already done that!) and add headers per the prescribed format.

Backfile conversion

Your backfile conversion utility will accomplish this for you. If you are currently doing business with the DoD, you are probably delivering your Design Disclosure Documentation via aperture cards. Your backfile conversion utility will take your scanned raster images, add MIL STD 1840 header information which it obtains from your aperture card holorith data (data printed at the top of aperture cards) plus metadata from your BIs, and structure your design disclosure data to meet the necessary requirements.

Remember that the type of electronic distribution that we are discussing at this point is the physical delivery of files per magnetic media or 'over the wire'. When we finally arrive at our destination, and implement true CITIS functionality, our customers, suppliers and other companies with which we conduct business will simply browse our PDM database as we do but under appropriate security limitations to access the required data. The middleman (electronic files in CALS type format) will not be 'delivered'. We will have 'delivery in place'.

On-line CCBs

Your on-line CCB is simply the integration of all of the above functionality into your PDM system plus the incorporation of the functionality of a 'workflow manager'. Your PDM will supply you with the capability to define and implement your present design change process, or better yet, the design change process presented in Chapter 4.

You simply need to work up some flow charts and then follow the instructions in your PDM manual for incorporating these workflows (or life cycles) into your PDM 'customization'. Test them out in your PDM test environment.

Give it a try. You will be surprised how easy it is.

Note: The integration of the tools described above into your PDM system is not as easy as 'falling off a log'. You will need personnel with expertise in this area in order to successfully accomplish the task. If you don't have these people currently on your payroll, you may want to consider retaining the services of businesses that do this for a living.

Phase 3

The functionality we implemented in Phase 2 allows us to capture and control design baselines electronically but we are still performing this activity only at that event which transitions the design data from informal control (WIP) to formal control, i.e., design review 3. Refer to Figure 2.1 for a refresher on the evolution of control. Remember that the desired state is when interface modules are available to integrate your PDM with your design tools (electrical and mechanical), and your design engineers and/or design draftspersons will be able to initiate their design task by simply signing on to the design network through your PDM and then launching the desired design tool through your PDM. Your designers will then work on their designs throughout the day (WIP states), and at the close of the business day, they will sign off the design tool, and the PDM will deposit the data created during the course of the day into the designers' personal work locations.

This process will be repeated until Design review 2 is completed. From then on, the 'workflow manager' application in your PDM will move your design files to the appropriate logical and physical vaults, and future changes will only be permitted after on-line approval from your designers' supervisor.

Baselines will be captured at design reviews by the PDM. They will not have to be manually captured by a CM person and the corresponding files moved back and forth to staging areas. These data will be entered once into the PDM database and updated by engineering personnel under the control of the PDM workflow manager.

Enter data once – use many times! You can imagine the savings in time and money. Sound like *déjà vu*? You're right. We discussed this concept in Chapter 4. Now, we will put it to use!

We are now ready to implement the functionality that will allow us to capture *all data* from the initiation of a design throughout the life cycle of the product without the intervention of CM personnel. They will continue to plan CM programmes, support CCBs, conduct CM audits, and resolve CM-related problems but the days of the old 'green eye shades', i.e., forms processing, database entry, and all other manual, labour-intensive CM tasks, will be over.

Integrate design tools

The first step in realizing the full functionality of your PDM system is to tightly integrate your PDM and third-party tools with your design tools. What do I mean when I say 'tightly integrate'? What's 'tight' versus 'loose' integration?

Tight integration is the condition whereby your design data, including references to part and symbol libraries, associations and linkages between parts, simulations, schematic diagrams, analyses, and all other data required to update or modify a design in the future are captured and controlled for each specific design as your data are generated by your design tools.

At this point in time, we are only able to accomplish this feat for 'flat files', i.e., those files which are two-dimensional in nature and do not have associations and linkages involved in their creation as do mechanical model tool generated designs.

At the time of writing this book, some of the primary mechanical tool suppliers have placed on the market their initial versions of PDM interface modules. Over the next several months other mechanical and electrical design tool suppliers will have their own versions of PDM interface modules available.

The initial goal of the suppliers mentioned above is to provide loose integration, i.e., the ability for a PDM system to launch design tools from the PDM and to capture individual mechanical and electrical designs. *Note*: We can encapsulate whole assemblies or projects at this point in time, and, for some design tools, individual designs, but in order to update a single design, we would have to check out and deliver these encapsulated files to the designers. They would subsequently have to open the encapsulated files, incorporate changes to the affected part, then encapsulate the designs once more and return them to the CM librarian for check-in to the PDM. This is a do-able workaround but very labour intensive and tedious.

The initial loosely integrated PDM interface modules will keep track of linkages and associations and allow us to capture, manage and later update individual designs. These initial PDM interface modules should also allow us to launch the design tools through our PDM systems.

So, what is missing? Well, the degree of automation required for the management of design data has to allow for integration activities such as allowing multiple design engineers to work on modifications to the same part at the same time with the PDM system tracking these changes and assuring that only the proper individual can and will review all proposed changes and approve only the correct combinations, i.e., total configuration management from the word go. This is truly 'tight' integration. It's on its way. In fact, some tightly integrated interface module 'beta' versions are currently being evaluated by industry.

Institutionalize WIP

As I said in Chapter 1, the introduction of our PDM systems into the WIP phase of our development programmes will be a culture change 'in spades' for many. One approach to minimize the impact is to implement the full functionality of our new PDM system one programme at a time. You can continue to capture baselines as designs are approved at the appropriate design reviews on most of your programmes, and, as new design and development activities are undertaken, you can start them off on the right foot by launching these new designs through your PDM systems, i.e., during WIP.

In any case, whether incrementally or all at once, you must force the necessary changes to the mindset of your employees in order to embrace the world of electronic data capture and control on an enterprise-wide basis.

Integrate strategic systems

Speaking of an enterprise-wide PDM, you should now start planning the continued expansion of the functionality of your PDM system to include some of the operations and activities currently controlled by your MRP and Consolidated Purchasing Systems (CPS).

There are certain things that MRP and CPS do quite well. They provide visibility into the impacts of changing lead times upon your master programme schedules. They translate engineering Design Disclosure Documentation into manufacturing planning and material procurement orders. They provide functionality that PDM systems do not duplicate and were not intended to provide.

It is therefore prudent and appropriate to partition the functionality expected of each of your systems according to best practices and tool capability. Once this planning has been accomplished, the necessary reallocation of data may be executed.

Integrate office tools

In order to edit and update documents other than design drawings and models, such as specifications, plans, reports, internal procedures and contract data, you will have to integrate your office tools, i.e., word processors, desktop publishing tools, presentation preparation tools and spreadsheets with your PDM system.

This process should not be problematic once you have integrated your design tools, since your office tools all generate flat files which are relatively simple when compared to your design models.

As-built process

Chapter 4 described the process for capturing and processing as-built configuration data. The better PDM systems make available (at additional cost) Product Configurator modules that provide the necessary functionality to automate this process and to embrace it within the PDM system scope of management and control.

The difficult part of the integration of this functionality is putting in place a uniform process to collect the appropriate data. Part scanning with wands for the capture of bar code serialization data is practical in some cases but everything from recording of the data on manufacturing planning sheets or manufacturing parts lists to visual capture and recording of the data by

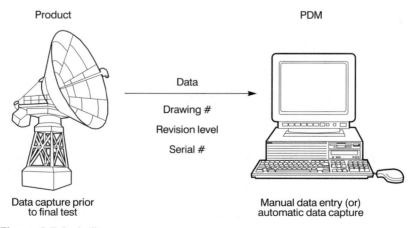

Product   PDM

Data

Drawing #

Revision level

Serial #

Data capture prior to final test   Manual data entry (or) automatic data capture

Figure 9.7 As-built process

configuration management personnel may be appropriate, depending upon many factors.

You will have to perform trade-off studies and experiments in order to know what is best for your company's manufacturing process. Figure 9.7 shows how the as-built process interfaces with your PDM system.

Phase 4

Phase 4 consists of the tight integration of your PDM system with your software development tools plus the establishment of CITIS functionality. You will already be managing your software CSA data with your PDM using the same methodologies employed by your hardware CM folks. However, you must transition from capturing baselined software code after it has been developed, i.e., as it goes under formal CM control ('issued' state) to WIP, as you did in Phase 3 for hardware.

Integrate software design tools

This evolution to WIP for software requires that you address issues similar to those which you addressed during your transition to WIP for your hardware development processes. Software design tool interface modules will have to be procured and evaluated, and the best ones selected and integrated into your PDM system. Institutionalization of the process must be achieved. In some ways, this effort should be easier, since you will have the advantage of 'lessons learned' from your venture into the world of hardware development during the WIP states.

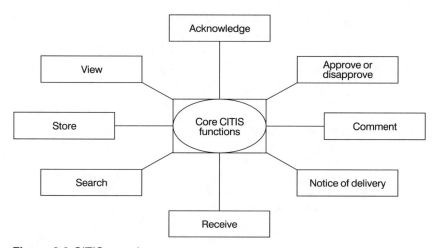

Figure 9.8 CITIS overview

CITIS

The only significant CITIS functionality missing from your PDM system at this point is your capability relating to communication with your external customers, vendors, suppliers and subcontractors.

Firewalls must be designed and implemented to protect the transfer of, and access to, data in each element of the PDM system (see Figure 9.8).

Once you have established a network with appropriate security provisions for the protection of your data, you will no longer have to deliver these data on magnetic media or over the wire to 'drop boxes' or directly to the addressee.

You will have full CITIS functionality.

Enter the data once – use it many times? That's the name of the game!

Integrate PDM functionality into programs

Each of the programs that you plan to integrate into your new PDM system must be evaluated for the appropriate scope and depth of integration. You may ask, 'why not integrate all programs totally?' and 'why even consider doing different things for different programs?'

My answer is, 'it's a trade-off between cost and the benefits accrued from bringing a program's data into the PDM "fold"'. You need to consider capturing and controlling current and future data *plus* decide what you are going to do about your legacy data. Your three most effective choices are:

1 Raster scan all existing drawings and induct the resulting images (and/or your legacy native CAD files) plus your design metadata into your PDM.
2 Create Business Items (BI) for your metadata and leave your drawings and other documents in their current state, i.e., paper and mylar drawings and other documents and aperture cards, and maintain them in their current locations. Point to them on your Data Item (DI) information screens.
3 Do nothing. Leave the old 'stuff' alone and just deal with new designs as far as your PDM is concerned.

You should separately consider each of these options for each of your programs. Chances are, when you evaluate cost versus benefits accrued, you will choose a mix of the above options.

Of course, you will capture and control electronic baselines for current programs or start right off capturing all design tool data during WIP for future programs. Keep in mind that it would be nice to be able to perform queries for all your data, past, present and future but we all have to consider the size of our wallets. One additional thing to consider is that, if you will be utilizing past designs for future projects or if you will be modifying past designs frequently for product enhancements, you may be better off scanning and managing your legacy data with your PDM.

User /data security

We have discussed the concept of user security occasionally throughout this book. However, I think it is appropriate to address it as a separate topic at this point in order to illustrate its importance to the successful management of your design data, internal procedures and contract data plus any other data you may induct into your PDM system.

The elements of user access security that you should consider are:

- Determine the level of clearance required by your contracts and programs, i.e., is ID and password sufficient or do you need to control access to your data on a 'need to know' basis?
- Can you define access control by state, i.e., WIP or released?
- Can you control access by vault (logical or physical)?
- Can you control access by type of PDM user, i.e., full PDM or casual view?
- Can you control access by business function, e.g., engineering, manufacturing, QA, management, product support, other?
- Can you control access by PC or workstation?
- Must you control access by government data classification, i.e., unclassified, confidential, secret, top secret?

Once you have identified your requirements by program you can decide the level of security access functionality required of your PDM system and incorporate that functionality accordingly.

Each PDM or casual view user must be validated initially (at least once) on your PDM. These personnel security access data will be stored in your PDM database for future reference each time they request access to data (see Figure 9.9).

Support

You will need to provide support to the users of your PDM system for the foreseeable future. The types of support required will be:

- Initial training for all types of users, e.g., CCB, designers, casual users;
- Initial user/PC/workstation setup;
- User role definition;
- Telephone 'help line' for problem resolution;

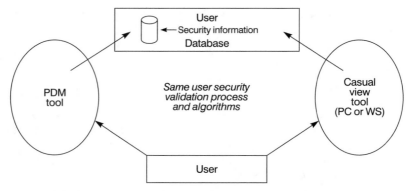

Figure 9.9 Security access overview

● Security access administration as your personnel transition in and out of your business and as they change jobs within your business.

You will be able to determine and fine tune the amount of PDM support required as you gain more experience with your PDM.

Networks

Figure 9.10 presents an overview of a typical PDM network configuration.

There are many network setups out there in the real world – everything from smoothly running 'well-oiled machines' to abysmal nightmares. The

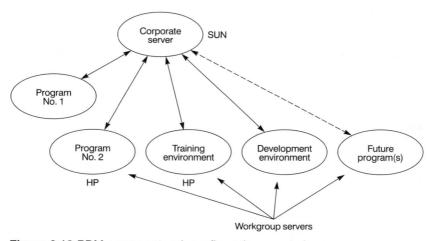

Figure 9.10 PDM server network configuration – overview

term 'spaghetti' is often used in describing real-world networks. I keep using the term real world because there is a mighty difference between what one usually sees on paper or in corporate maintenance manuals and what is really 'out there'. The inside scoop is often only known by the 'boys in the box', i.e., those network administrators who, by the grace of God, know how to keep you and your people on-line most of the time.

You will want to install your PDM as a distributed system, as mentioned above, to allow for expansion and to facilitate the best throughput and performance. Therefore, it is an appropriate activity to evaluate your real-world network(s) before you install your PDM system in order to preclude problems down the road.

Summary

Now you have it! Your PDM system is installed, and you have integrated its functionality into your current programs and some or all of your mature, legacy programs.

After you have been using it for a while, you can start to reduce your CM, Drafting and Document Control Centre workforce by appropriate (but significant) numbers. You will see a dramatic reduction in your business cycle times and costs. You will deliver products whose designs are better identified and controlled. You will do all this in compliance with your customers' requirements and your company's guidelines.

Your customers and suppliers will be thrilled with the improvements in communication and access to program data.

You have now reached your goal in your journey towards excellence in CM. It didn't take so long after all now, did it?

Congratulations!

10

The CM templates

This chapter provides a series of templates which, when filled in, will form the overall template set for your CM/PDM system. Table 10.1 lists the CM template set.

Table 10.1 CM template set

Title	Figure number
Tailoring and planning template	10.1
Configuration Identification template	10.2
Configuration Control template	10.3
Configuration Status Accounting template	10.4
Configuration Audits template	10.5
Transition to production template	10.6
Production and support template	10.7
Software and firmware template	10.8
Problems and resolutions template	10.9
Automated CM template	10.10

These templates, plus two additional templates contained in Appendix C, address all CM and PDM activities covered in this book. They will provide you with the opportunity to perform your CM/DM/PDM planning, execute your plans and register the degree of compliance or non-compliance as you proceed with the implementation of your PDM system.

Software copies of the templates are provided on the encosed CD ROM. I recommend that you update them periodically as you establish the CM 'best practices' in your operations and implement the functionality of your new electronic PDM in your company.

The intended meaning for the heading categories for each template element to be addressed before, during and after you have implemented your automated CM system is as follows:

- *'Ownership'* – Have you planned this template element? Has ownership been assigned? Are you sure you have 'buy-in'?
- *'Action'* – What is the next action that must be taken to drive this template element toward a satisfactory conclusion?

- *'Status'* – Has this template element been started? Has it been clearly communicated to all participating organizations? Has it been completed?

Please take the time to evaluate your true condition before you start this project. Establish a CM/DM/PDM implementation status baseline. Be excruciatingly truthful in your responses. Only then will you be able to plan effectively for the actions you must take as you begin your journey to excellence in CM.

I also suggest that you complete a new set of templates as you complete each of the four phases of functionality implementation for your new PDM. You will be surprised how the lessons learned during each preceding phase will aid your planning for the next phase.

Good luck!

CM element	Ownership	Action	Status
Contract requirements	————	————	————
CM program negotiation results	————	————	————
CM program milestone charts	————	————	————
State chart strawman	————	————	————
Configuration Identification			
Baseline capture plan	————	————	————
Design family tree	————	————	————
Design listings and design file definitions	————	————	————
Configuration Control			
Design review schedule	————	————	————
CCB membership	————	————	————
MRB membership	————	————	————
PRB membership	————	————	————
Configuration Status Accounting			
Data capture process	————	————	————
Data processing	————	————	————
Data reporting process	————	————	————
Configuration Audits			
Audit plans	————	————	————
Audit schedules	————	————	————
Audit data capture methodology	————	————	————
Configuration Plan			
Core elements	————	————	————
Program-unique information	————	————	————
Internal CM procedures to reference	————	————	————

Figure 10.1 Tailoring and planning template

CM element	Ownership	Action	Status
Design identification scenario	_____	_____	_____
Configuration Item selection	_____	_____	_____
Serialization plan	_____	_____	_____
Part marking plan	_____	_____	_____
Design review checklist	_____	_____	_____
Baseline capture plan			
Functional	_____	_____	_____
Allocated	_____	_____	_____
Development	_____	_____	_____
Product	_____	_____	_____
Technical data package			
Deliverable data	_____	_____	_____
Non-deliverable data	_____	_____	_____
Tactical data	_____	_____	_____
Non-tactical data	_____	_____	_____

Figure 10.2 Configuration Identification template

CM element	Ownership	Action	Status
CCB			
Process flow chart	_____	_____	_____
Membership	_____	_____	_____
Start date or event	_____	_____	_____
Forms or PDM screens	_____	_____	_____
PRB			
Process flow chart	_____	_____	_____
Membership	_____	_____	_____
Start date or event	_____	_____	_____
Forms or PDM screens	_____	_____	_____
MRB			
Process flow chart	_____	_____	_____
Membership	_____	_____	_____
Start date or event	_____	_____	_____
Forms or PDM screens	_____	_____	_____
PDM			
Tools – software	_____	_____	_____
Staging areas	_____	_____	_____
Vaults	_____	_____	_____
WIP – workflow	_____	_____	_____
CCB – workflow	_____	_____	_____

Figure 10.3 Configuration Control template

CM element	Ownership	Action	Status
Database established	————	————	————
Data capture scenario	————	————	————
Data processing	————	————	————
Data reporting			
• Deliverable	————	————	————
• Non-deliverable	————	————	————
PDM			
Query reporting screens	————	————	————
Data file delivery format	————	————	————
• IGES	————	————	————
• CGM	————	————	————
• PDF	————	————	————
• Raster	————	————	————
• TIFF	————	————	————
• HPGL	————	————	————
• STEP	————	————	————

Figure 10.4 Configuration Status Accounting template

CM element	Ownership	Action	Status
FCA			
FCA plan	_____	_____	_____
Design evaluation tests	_____	_____	_____
Qualification tests	_____	_____	_____
Theoretical error analysis			
FCA report	_____	_____	_____
PCA			
PCA plan	_____	_____	_____
Data capture			
• FAI/revision/serial #	_____	_____	_____
Data processing	_____	_____	_____
PCA report	_____	_____	_____
Action item closure	_____	_____	_____
Product baseline established	_____	_____	_____
CVA			
Data capture	_____	_____	_____
Data processing	_____	_____	_____
Data reporting	_____	_____	_____

Figure 10.5 Configuration Audits template

CM element	Ownership	Action	Status
Transition team			
Membership	_____	_____	_____
Start date or event	_____	_____	_____
Program schedule	_____	_____	_____
Action item database	_____	_____	_____
Database verification			
CSA	_____	_____	_____
MRP	_____	_____	_____
CPS	_____	_____	_____
PDM	_____	_____	_____
Material procurement			
Vendor coordination	_____	_____	_____
Design change cut-in status	_____	_____	_____
Design file verification	_____	_____	_____
Design release to manufacturing	_____	_____	_____
Customer involvement			
Design reviews	_____	_____	_____
Design approval	_____	_____	_____
Material orders	_____	_____	_____
Change approval	_____	_____	_____

Figure 10.6 Transition to production template

CM element	Ownership	Action	Status
CM plan	_____	_____	_____
CCBs – hardware and software			
• Internal	_____	_____	_____
• External	_____	_____	_____
Customer interface	_____	_____	_____
Major change process	_____	_____	_____
MRB	_____	_____	_____
PRB	_____	_____	_____
CVR			
• Process	_____	_____	_____
• Schedule	_____	_____	_____

Figure 10.7 Production and support template

CM element	Ownership	Action	Status
Software CM plan	_____	_____	_____
Software build process	_____	_____	_____
CM milestones	_____	_____	_____
Design review schedule	_____	_____	_____
CM checklist	_____	_____	_____
Hardware/software coordination			
• Planning	_____	_____	_____
• Communication	_____	_____	_____
• Scenario	_____	_____	_____
• Lessons learned	_____	_____	_____
Firmware CM processes	_____	_____	_____
PRB			
• Communication	_____	_____	_____

Figure 10.8 Software and firmware template

CM element	Ownership	Action	Status
PRB			
• Membership			
– Chairperson	_____	_____	_____
– Administrator	_____	_____	_____
– Project engineer	_____	_____	_____
– Representatives	_____	_____	_____
• Process	_____	_____	_____
– Database in place	_____	_____	_____
– Problem reporting	_____	_____	_____
– Categories defined	_____	_____	_____
– Corrective action	_____	_____	_____
– Problem assignment	_____	_____	_____
– Status reporting	_____	_____	_____
– Closure verification	_____	_____	_____
• CCB Communication	_____	_____	_____
– Interfaces	_____	_____	_____
– Feedback	_____	_____	_____

Figure 10.9 Problems and resolutions template

CM element	Ownership	Action	Status
Planning			
CM process survey	————	————	————
CM process 'tune-up'	————	————	————
PDM project engineer selected	————	————	————
PDM team selected	————	————	————
Requirements analysis	————	————	————
System specification	————	————	————
• Data model – objects	————	————	————
• Data model – attributes	————	————	————
• Workflows	————	————	————
PDM tool evaluation plan	————	————	————
PDM vendor evaluation plan	————	————	————
Implementation strategy	————	————	————
Functionality verification plan	————	————	————
Documentation plan	————	————	————
Training plan	————	————	————
• Overview	————	————	————
• Detailed	————	————	————
Resources identification	————	————	————
Program integration plan	————	————	————
Project schedule	————	————	————
Project progress reporting scenario	————	————	————
Metrics capture plan	————	————	————
• Baseline before PDM	————	————	————
• After PDM in place	————	————	————
Project funding	————	————	————

Figure 10.10 Automated CM template

CM element	Ownership	Action	Status
Functionality implementation			
Phase 1			
• Check-in/checkout	_____	_____	_____
Phase 2			
• View/markup	_____	_____	_____
• Raster edit	_____	_____	_____
• Distribution	_____	_____	_____
• Backfile conversion	_____	_____	_____
• On-line CCB	_____	_____	_____
Phase 3			
• Integrate design tools	_____	_____	_____
• Integrate office tools	_____	_____	_____
• Integrate MRP tools	_____	_____	_____
• Institutionalize WIP	_____	_____	_____
• Establish automated 'as-built' process	_____	_____	_____
Phase 4			
• Integrate software tools	_____	_____	_____
• Integrate financial tools	_____	_____	_____
• Establish CITIS network and firewalls	_____	_____	_____
• Establish supplier interfaces	_____	_____	_____
Support			
Telephone 'help desk'	_____	_____	_____
Network	_____	_____	_____
Process administration	_____	_____	_____
• PDM tool	_____	_____	_____
• User add/delete/modify	_____	_____	_____
• Security	_____	_____	_____

Figure 10.10 Continued

Appendix A
CMP preparation guidelines

Introduction

Roadmap

Your Configuration Management Plan (CMP) will serve as the roadmap for your configuration management programme. From the time you embark on your initial proposal activities to the time you initiate development activities, you will be planning the configuration management processes that you will put into place, and you will be negotiating these processes with both your internal functions and your customer(s). The results of these planning activities and negotiations will be documented in your CMP. Then, from the time you start your requirements analysis and design definition throughout development, test, transition to production, production, deployment, delivery, on to the end of your maintenance and support activities, your CMP will guide you, your employees and your customers along the way.

Your CMP should tell the story, in clear and simple language, about how you have agreed to control your design data during development, capture and control your development and production baselines, conduct your FCA (prove that your design works) and PCA (prove that you can build your product), control the ordering of your materials, provide status information (internally and externally), control the fabrication and build of your hardware and software, run your CCBs, MRBs, PRBs, design reviews, transition team meetings, plus perform periodic audits to assure compliance to your documented concurrent engineering and production processes.

Your CMP will also serve as a reminder of those agreements made early on in the programme, just in case someone forgets.

As you progress through this appendix, you will notice that I have created three separate categories. These categories provide different options and guidelines for different sized businesses.

Small business
Medium-size business
Large corporation

The purpose of this categorization is to make it easy for you to focus on the information you need to know to generate a meaningful and useful CMP appropriate for the size of your business and the specific programme involved.

Whether you operate a small, medium or large business, your single most important CM tool is your CMP. Even if you don't design the products you are building, you need a CMP to control the design provided to you by your customer or engineering subcontractor.

Generic CMP versus programme-unique CMP

I wish I had a dollar for every time I've been asked to write a 'generic' CMP. You would think a CMP was the same as a procedure for changing the tyre on a bicycle, i.e., pretty much the same for all bikes. By the way, this is a good time to emphasize the difference between plans and procedures. I will do that as soon as I complete this section on generic CMPs.

A generic plan is just that – generic: a plan for all occasions! That is not possible with CMPs!

Your CMP documents the merging of your 'best CM practices' and the customer's requirements for a given programme. The CM processes negotiated for your programme represents trade-offs between levels of control and cost. These processes are documented in your programme CMP. If military standards are invoked for a programme, your CMP must document how the requirements specified therein are to be satisfied.

Therefore, you must bite the bullet and be prepared to address the specific CM requirements of each new programme as it comes along. You may, by all means, have standardized CM 'best practices' documented in your internal, step-by-step procedures (you should base them on the processes provided in this book). You will use these 'best practices' plus the templates provided in Chapter 10 to plan and negotiate your CM programme.

CMPs versus CM procedures

I don't want to belabour this point but it deserves coverage here. Plans document relatively high-level decisions, provide schedules to identify milestones and establish completion criteria, reference requirements and related documents, and describe, in general terms, how you plan to conduct business for a given programme. Enough information should be provided in a plan to allow the reader to understand, to a reasonable degree, the interrelationships between processes and programme driven events, including a substantive description of roles and responsibilities of those individuals and functional areas performing on the programme.

Procedures, on the other hand, provide detailed, step-by-step, instructions as to how you are going to implement these processes. You should be able to use identical or similar procedures from one programme to the next. Your procedures should represent your 'best CM practices', i.e., they should represent your implementation of the contents of this book.

CMP template

Your CMP template is the set of documents that will guide you through your CM planning activities for current and future programmes. This set of documents consists of:

1 The templates provided in Chapter 10 and Appendix C (including the notes and excerpts from this book that you will copy into the comments section of these templates).
2 The figures provided throughout this book
3 The marked up sets of these templates that you will generate as you progress towards your goal of excellence in CM.

Now that you have the gist of the process, you must study the information I have provided at least one more time (re-read the book) and take notes. Hopefully, you will want to refer to the chapters contained herein many times in the future. I have provided a form for your use for jotting down notes in Figure A.1. I think it will make things a bit easier for you, at least at first. As I have said before, this process will not be as easy as falling off a log. Persevere, and you will be rewarded. You have no other choice in today's business climate.

CMP length

Your CMPs should not be voluminous documents. Nobody will read them if they are. The length will vary from programme to programme but a good rule of thumb is no more than fifty pages (thirty pages is good), not including appendices which contain definitions, acronyms, special topics and forms.

Be concise but don't skimp on the necessary information. A few well designed flow charts such as your process for design baseline capture and change control from beginning to end will go a long way to convey the information you need to communicate. Use the flow charts in this book and Figure 2.1 to get yourself started. You should also include your organizational reporting chart. Your customers will want to know who's the boss and who's responsible for the various aspects of your business.

Remember that, whenever a question arises during any phase of your programme, you want your people to automatically go first to their best source of reliable information – their CMP. If you do a poor job in writing your CMP, they will come to you, instead, every time!

Program _____

CM representative _____

Chapter	Topic	Points to remember	Reference

Program _____

CM representative _____

General notes and observations:

Figure A.1 Notes for the CM template

Tailoring and planning

Communication with engineering

In this section, we will first discuss the importance of communicating the details of the job to be done with those who are going to help you do it. You should tell them about the customer's requirements and discuss how you will propose your 'best CM practices' to your customer, tailored to meet their needs, yet not cause too much damage to their wallet.

You must then make sure that all of your key personnel understand the content of the work, the interrelationship of CM and the engineering, quality assurance and production processes and agree to the roles and responsibilities of all involved during the various stages of the programme. You will then proceed to document the results of your planning and tailoring in your CMP. Place this information right up front in your CMP. It is very important that your customer knows that you have done a good job in the planning and tailoring area. They can help you iron out details later but you must demonstrate that you have first planned a solid foundation for your (and their) programme.

Start with your engineering personnel because it is they who are involved most during the early stages of the programme. Understand that you need to involve other functions, such as producibility engineering, components engineering, quality assurance, manufacturing, product support, and sourcing in both primary and supporting roles.

If you operate a small business and your engineering department consists of one individual, get them involved at this time.

Obtaining 'buy-in'

The key to success for a smoothly running CM operation is the 'buy-in' of your CM/DM/PDM system users and, especially, of those key individuals who will help you integrate your PDM system into your engineering, quality assurance, manufacturing and CM disciplines and methodologies.

Without their buy-in, you don't have a prayer for success. Actually, today it is much easier to garner that cooperation than it would have been two or three years ago. Engineers and draftspersons know that they need help in managing their design data files. Quality assurance and production personnel want to get rid of all that paper. Fortunately, you are operating today on the winning side of the curve.

In order to assure buy-in of these important elements in your operation, you have to be 'up front' with your people and let them know, as early as possible, what is going on.

I recommend a series of 'CM/DM/PDM overview' discussions where you describe in sufficient detail for thorough understanding just what your CM requirements are and how you plan to achieve them. Then you should describe where you are going with your PDM system and tell your people how you plan to get there.

Open yourself up to their questions and ask for their help. You will need it, and they will be glad to provide it. After all, they have a stake in your programme's success, too.

Don't forget to document the results of these discussion and planning sessions in your CMP.

Roles and responsibilities

As you and the representatives of your functional areas discuss and complete the planning checklists provided herein, the roles and responsibilities of your personnel will become obvious. Document these roles and responsibilities in your CMP.

Checklists and templates

The checklists and templates presented in this book provide the basis for what you and your organization need to generate a good CMP. The Design Review Checklist (Figure 3.1), the CM Planning Schedule (Figure 3.2), the CM Planning Checklist (Figure 3.3), and the ten CM Planning Templates provided in Chapter 10 comprise the checklist and template set that you will need to plan and execute your CM programme and generate your CMP.

The order in which you fill out the checklists and planning sheets doesn't matter. I advise you to read them all once, then fill out what you can. You should then search out information you don't have at your fingertips. This activity will provide you with the opportunity to talk to those individuals in your organization that I have spoken of earlier in this appendix.

Once you have completed your first draft of the planning sheet/template set, pass it out for review. Then call a meeting and discuss each entry on every sheet. You will be amazed at the amount of good information that will result from this meeting. It may be tedious but it is worth it. This time invested up front will reward you manyfold as your programme progresses.

After you have secured agreements and revised your set of checklists and templates, you should edit them at least one more time and include them in an appendix to your CMP. You will see that these checklists and templates describe not only your best CM practices but document the unique processes that you intend to apply to the programme being planned.

You should periodically update these checklists and templates. They will provide an excellent source of 'lessons learned' for future programmes and projects.

Military standards and commercial standards

I touched briefly on the subject of military standards (MIL STD) and commercial standards in Chapters 3 and 4. The Procurement Reform

Initiative, i.e., the Perry Initiative, undertaken by the DoD is, in part, designed to reduce the number of MIL STDs utilized by the DoD to procure products. Some MIL STDs, though, such as the Interface Specification, MIL STD 2549, will be a mandatory requirement for most DoD suppliers and contractors.

Commercial standards and guidelines may, on the other hand, be recommended by DoD suppliers as the way in which they propose to run programmes for either the DoD or for commercial customers. The government can then decide whether or not it can live with the processes described in these commercial standards.

Some companies are in the process of generating their own internal 'best practice' procedures, based upon commercial standards and guidelines, military standards, ISO guidelines and requirements, where applicable, and just plain good sense. You need to become aware of the 'playing field' on which you will be conducting your programmes. Make sure you maintain compliance to those standards which your customer requires and generate your CMP accordingly, using the 'best CM practice' techniques taught in this book.

Risk management

As a final note in this section about tailoring and planning in preparation for the documentation of your CM processes in your CMP, we should introduce the concept of risk management. After all, that is what CM is all about.

CM is a trade-off between control and cost. The amount of CM you can afford depends upon the level of risk you are prepared to take and the size of your customer's wallet. We still have to deal with what's real! More control and traceability costs more money up front but pays off in costs saved down the line. I don't believe it is appropriate to try to sell you on the idea of putting lots of CM procedures in place at this time. If you haven't got the idea of the value of a solid CM programme by now, then I've done a pretty poor job of writing this book.

On the other hand, you should be aware that there is, indeed, a point where you must decide upon the appropriate amount of CM for each of your programmes.

This is particularly true for small- and medium-size businesses. You can start with a good CMP and continue to run your CM programme in a controlled environment without installing a sophisticated PDM system and still reduce your risks and enjoy the resulting savings.

Use the 'best CM practices' defined herein and integrate them into your programme as appropriate to meet your customer's requirements within their budget. Some decisions may be hard but you have to make those decisions in as well informed a position as possible. By now, you should be very well informed about the values of CM and the methods required to implement an efficient and effective CM programme.

Configuration Control Boards

One of the subjects that is always covered in CMPs is the topic of Configuration Control Boards (CCB). The reason for this is simple. CCBs are the vehicles by which we control changes to our designs after we have captured our design baselines.

The reason we convene CCBs is to tap the resources that know the most about the fundamental ways in which we conduct business. As I have pointed out throughout this book, it is imperative that the appropriate people review proposed changes to our designs.

Small and medium-size businesses may have limited functions and personnel to draw from for their CCB membership. This is OK. The important thing is to involve those individuals whom you can trust to carefully evaluate proposed changes and make wise business decisions.

With such a significant amount of responsibility riding upon the shoulders of the CCB members, it is prudent to define roles and responsibilities of the various CCBs as early in your programmes as possible. Actually, in most DoD programmes, the CMP is a deliverable item, and this fact forces the CCB architecture and composition decisions to be made in a timely manner.

Internal CCBs

Internal CCBs should be organized by the categories of hardware CCBs and software CCBs. Sometimes you will see integrated product development 'team' (or worker level) CCBs plus internal programme level CCBs. The latter is the exception rather than the rule, though. The programme level CCBs generally encompass your customer representatives and are usually put into place after the product baseline is established. Figure 6.1 shows the recommended CCB heirarchy.

Your CMP should provide an illustration similar to Figure 6.1 to show the organizational hierarchy of your internal CCBs. It should also show the relationship between your internal and external CCBs.

You should identify the roles and responsibilities of your required and supporting CCB members as described in Chapter 4. You should also define in sufficient detail the process by which proposed changes to your designs are presented to your CCBs, the manner in which they are reviewed and approved or disapproved, the process by which your designs are updated and the mechanisms you have instituted to assure proper cut-in of design changes to your hardware and software products.

It is perfectly acceptable to point to other internal procedures for detailed descriptions of exactly how these processes are performed. Your CMP should, however, provide a roadmap that the careful reader can follow to understand the overall process.

You will need to discuss how information and action items are communicated between your internal CCBs. This communication line is essential in assuring that important items do not get dropped in the crack. The details of this process were presented in Chapter 4 and Chapter 6. My purpose here is chiefly to advise you as to the proper content of your CMP and to drive home the fact that you must show, in your CMP, how you communicate, internally.

External CCBs

Following the establishment of your product baseline, your customer will probably want to sit in on your CCB to review and approve changes (at least Major (Class 1) proposed changes), or to conduct a programme level CCB of their own, especially if your contract is with the DoD. I do believe, however, from my observations over the past few years, that even our military customers are leaning towards control of our designs by controlling our design specifications and leaving the control of our individual detailed designs to ourselves, i.e., the contractors.

In any case, you should negotiate this facet of the change review and approval scenario with your customer and try to come to an agreement that will provide the most 'bang for the buck' when it comes to the level of activity assigned to your external CCB. Your customer must understand that here, as in other areas of CM, it is primarily a matter of cost versus control. You and they must answer the question, 'how much CM do I need'? If you can convince your customer that your CM methodology is properly integrated into your internal design and development processes and that you have a solid CM process in place during production, you may be blessed with minimal customer participation and intervention during the latter stages of your contract. This is one area where your planning activities with your customer early in the programme will pay off as demonstrated by the amount of trust they are prepared to place in you and your processes at this stage of your programme.

Small businesses may never see the need for an external CCB. Your process will be more streamlined if you aren't required to obtain customer approval for proposed changes.

CM processes

Now that you have the planning, tailoring and organizational matters thoroughly covered in your CMP, it is time to get down to the specifics as to exactly how you are going to run your CM programme for this contract. You should present and document your CM processes in the same categories that they were discussed in the main part of this book:

Note: The detailed descriptions of the following processes and the reasons for their application within your programmes are presented in Chapter 4 and are also summarized in the templates provided in Chapter 10:

- Configuration Identification;
- Configuration Control;
- Configuration Status Accounting;
- Configuration Audits.

Your task in generating your CMP is to describe, in plain language, how you intend to address the areas of activity presented to you in this book, simple as that! Remember, if it's true, it has to be simple!

You also need to document the answers to the following questions as to why you will be performing the CM job for each category described above:

- What are the reasons why I should perform this activity?
- What will happen if I don't do it?
- What is the process that I will follow in order to perform this task?
- What are the risks that I am taking?
- What are the 'red flags' that will tell me if I am doing something wrong?
- What is the corrective action that I can take to remedy a bad situation if it occurs?

These questions were posed by the Willoughby templates (1985, 1986). I have borrowed them to pass on to you because they are the right questions to ask. I do, however, want to make sure that Willoughby receives the proper credit for them. He certainly has the right thinking in this area.

Automated CM

You will have achieved fully automated CM when your new PDM system becomes tightly integrated into your CM and engineering development tools and processes. You will need to cover this topic in your CMP.

Small businesses who do not plan to implement the functionality of a PDM system can, of course, skip this section.

Small and medium-size businesses that are considering implementing a limited scope PDM such as a local area network (LAN) should discuss this activity in their CMP, even if the programme is going to be initiated via a paper-based scenario.

Large corporations, however, should cover this topic with a comprehensive section in the CMP. Your customer will want to know the extent to which you are automated, and your personnel will need the roadmap to follow in performing their CM tasks.

You should be very careful to explain, in simple terminology, your 'end condition' of automated CM, i.e., paint a picture of where your are now and articulate how you are going to reach that blissful final state by increasing your PDM functionality a step at a time and by simultaneously integrating your new PDM functionality into your programmes and processes. Of course, by the time you need to generate your next CMP, you may already be at or beyond the final phase of the customization and integration of your PDM system.

Chapter 9 provides the information you will need to both implement your new PDM system and to document its functionality in your CMP. You may use the figures contained in Chapter 9, as well as those provided throughout this book, as the basis for similar illustrations for your own CMP.

You should provide in an appendix to your CMP, or incorporate into a separate internal procedure, the screens to be used by your PDM system and the procedures required to operate the system. Don't forget to negotiate up front the resources required for the use of your system (hardware, software and network hookups). Also, plan for training PDM system users, including your customers and suppliers.

Finally, document it all in your CMP.

Good luck on your CM programmes!

References

Willoughby, Willis J. Jr, (1985) *Transition from Development to Production*, Willoughby Templates, DOD 4245.7-M, Department of Defense.

Willoughby, Willis J. Jr, (1986) *Best Practices: How to Avoid Surprises in the World's Most Complicated Technical Process*, Willoughby Templates, NAVSO P-6071, Department of the Navy.

Appendix B
Questions and answers

Introduction

Several readers who read pre-publication copies of this book have asked questions relating to the topics covered herein. Some of these questions required rather extensive answers in order to be complete enough to be of any real value and to tell the inquirer what they wanted to know. I was originally planning on incorporating the content of these answers into the text of the book but, upon reflection, reconsidered. I figured that, if I tried to blend the information contained in my answers to reader's questions into the text, I would quite possibly disrupt the natural flow of the original text and perhaps do more harm than good.

I have, therefore, decided to dedicate an appendix to the task of providing this information in such a manner that you will receive additional, valuable information without becoming confused by 'overstuffed' paragraphs in the text of the book. After all, that is what appendices are supposed to be for, isn't it?

The following list of question topics is provided to guide the reader to topics of interest:

Question 1   Clarify engineering prototype
Question 2   Explain 'encapsulation'
Question 3   Explain Business Items and Data Items in more detail
Question 4   What is Figure 4.1 trying to show?
Question 5   Clarify your definition of CIs
Question 6   MRPII systems – track doc # and rev?
Question 7   Explain IGES and raster formats
Question 8   Still need Configuration Audits?
Question 9   Why have a transition to production period?
Question 10   Effects of Perry Initiative?
Question 11   Identify PDM cost savings
Question 12   How to sell PDM to management?
Question 13   What would I do different?
Question 14   Requirements Specification template?
Question 15   Still need effectivity information on original documentation?
Question 16   Workarounds for 'tight integration' – explain in greater detail

Questions and answers

I will follow a question and answer format starting with the reader's question.

Question 1

Can you clarify the use of the engineering prototype and explain its relationship to the FCA and PCA? Also, you say, 'FCA is conducted on the engineering prototype. PCA is conducted on the first production unit'. How many companies are really doing it this way? In a programme I work on, the PCA was conducted on the third production unit, and the FCA was conducted in phases and lasted a few years. Guess what happened!

Answer 1

The engineering prototype is used as the test vehicle for all tests that make up the FCA, i.e., design evaluation tests, qualification tests – whatever tests are necessary to demonstrate to yourselves (and to your customer, if applicable) that you have met the requirements defined in your system level specification(s). Your prototype represents your design (functionally) and has been updated as changes have been made to your design during its construction. At this point in your development process, a certain amount of 'goodness' is expected in your designs (as verified by your design reviews), and your prototype hardware/software should agree with your design documentation (paper or electronic). If your designers did their jobs perfectly (rarely the case), your testing would confirm that the design works. In the real world, however, you have probably found and corrected deficiencies in your design as you built your prototype and updated both your design documentation and your prototype hardware and software. Note that the only validation here is to the functional requirements.

The only thing you have left to prove in order to establish your product baseline is that you can build additional copies of your prototype (design) in a production environment, and so you perform your PCA on the first production unit to demonstrate that:

1 Your manufacturing planning (based upon your engineering drawings) is good enough to build production units.
2 This unit's measurements and physical characteristics match those defined on your engineering drawings (sometimes called First Article Inspection – which may include 100 per cent inspection, or something less).
3 This unit is built to the correct revision level and its assemblies and parts are properly serialized.

Finally, an Acceptance Test (subset of your qualification/design evaluation tests) proves that the production unit is functionally the same as your prototype unit.

You could perform your PCA on a second, third or fourth production unit but this would only delay the declaration of your product baseline, a major point in the evolution of your Technical Data Package (TDP) and in your release of your design for material procurement and full-scale production. For military contracts, this is also the point where the designs (drawings or electronic data) are submitted to your military customers for authentication. From this point on, they get involved in changes to your designs. This is true for the general case. There is a trend today for military customers to leave detailed design change approval to the contractor. The customer controls the overall design by controlling changes to the top-level specifications.

Question 2

Explain the term 'encapsulation' as it relates to PDM capture of design files.

Answer 2

Encapsulation is the process of enclosing multiple files within a single file, directory or 'folder' with a single filename. For example, mechanical and electrical design modelling tools produce not only (two-dimensional or 'flat') drawing images but generate pointers, linkages and associations (all separate files) to other parts, part and symbol libraries, etc. Analyses and simulations are also related to these designs and are required in order to re-use or modify these designs in the future. All this 'stuff' must be captured in a manner in which it can be inducted into a controlled vault and recovered in the future and re-installed on a similar design tool to re-use or modify the design.

So, we 'encapsulate' these designs. Think of the process as wrapping all these files up together and tying a ribbon around them. We hang a tag on the ribbon and write a unique filename on the tag. Then we check the whole thing into the PDM.

Various processes are available to accomplish this. I use the 'tar' process for mechanical designs and the 'zip' process for electrical designs (to accommodate the longer filenames and path definitions).

Question 3

Explain Business Items (BI) and Data Items (DI) in more detail.

Answer 3

The PDMs I discuss in this book are 'object-oriented'. Business Items (BI) and Data Items (DI) are objects. BI are objects that contain 'metadata'.

Metadata is information (such as the information about a drawing, e.g., title, revision, size, number of sheets, programme and product used on and distribution information). It can be information about hardware parts or software code. Any kind of information can be represented by, and contained in, a BI. The key point to remember about BIs is that they represent information – not physical things. A DI, on the other hand, is an object that represents and is used to control physical things such as files or paper. We capture files, check them into our PDM and control them. We point to paper in those cases where we do not plan to raster scan the paper or mylar drawings or aperture cards and capture the resulting files in our PDM system.

There is a relationship between BI and DI. For example, for designs, we create a 'design document' BI to contain the information about a part's design. We create DIs for the files that contain the native and neutral files that describe the part (usually one file per sheet of the drawing). We attach (relate) the DIs to the BI. This relationship is also a BI object. When we want to modify the design, we check out the DIs, modify the files, check them back into the PDM and update the revision level specified in the BI.

This is a very simplistic explanation. We also attach Change Notice BIs, Change Notice Disposition sheet BIs and problem report BIs to the design document BI (drawing information) for the CCB members to review. Life cycle (workflow) information is also represented by BIs. I suppose I could write a whole separate book about the theory used to generate the data model (including BI and DI and workflows) for the PDM system. However, I don't want to saturate my readers.

Question 4

Clarify what Figure 4.1 is trying to show.

Answer 4

Figure 4.1 is meant to show the initial implementation of the check-in and check-out functionality of our PDM system. We initially capture files and check them into our PDM by having engineering or drafting move them to the staging server. A CM librarian then moves these files to their personal 'work location' (like a personal vault) and then checks these files into the PDM.

After a proposed change has been approved by the CCB, the CM or drafting librarian checks the files to be modified out of the PDM to the staging server. Drafting then moves them to the original design tool and performs the necessary updates. The process is then reversed, and the files are checked back into the PDM.

Note that this is only a temporary workaround until we have 'tight integration' between the PDM and our design and office tools. Then, the user (designer, draftsperson, office worker) will launch their design or office tools through the PDM and generate data which the PDM will capture. The data goes into the PDM once and stays there. It will be accessed, under PDM security limitations, by all personnel who have the need and authorization to use it (engineering, drafting, CM, manufacturing, quality assurance, field service, the 'ilities', sourcing, vendors, customers, etc.). 'Enter once' – use many times. That's the goal!

Question 5

In your discussion on Configuration Items (CI) you say that 'every part with a drawing number assigned by the cognizant company, is a TCI (Technical Configuration Item)'. I don't quite agree with this. You also say, 'CCI are line items specified in a contract'. A definite *yes* from a contractor point of view, but I think the customer is not going to agree with that. Can you elaborate on your definitions of CIs?

Answer 5

You are not the only person to disagree with me on the subject of 'what is a CI?'. There was a lengthy thread on this topic several months ago on the CM Working Group topic list. If you are interested in the details, I will forward the related messages to you (I am a packrat – never delete old files until I run out of room). Basically, what I am trying to say is simple, though:

1 Different folks have different ideas about what a CI is. I tried to identify those different CI definitions in my book.
2 I believe that a CI is a design that you want to capture and control. If you have gone to the trouble of assigning a number to it, then you should capture (baseline) it and control it. You may also want to report status on it and perform audits on it. Sounds like Configuration Management to me, i.e., management of a CI.

Question 6

I don't know of a MRPII system that can use or track document number and revision level. Does such system really exist?

Answer 6

Our MRPII system (PIOS – Production Ordering and Inventory System) is currently our Record of Authority (ROA) by default. Our old engineering documentation control system (RESPONDS – only the 'ancients' knew what it stood for) was replaced by PIOS. As a result, we needed a place to store our revision status information. Now that we have a new PDM, we will be transitioning the Record of Authority to the PDM. We still have to resolve the interface between our PDM system and PIOS (engineering BOM versus manufacturing BOM, etc.).

Question 7

Could you explain IGES format and RASTER format, and the difference between the two?

Answer 7

IGES and raster files are two different types of 'neutral' format files (as compared to CAD-generated 'native' files). They are used primarily for viewing and distributing electronic versions of engineering drawings and other documents. IGES (Initial Graphics Exchange Specification) files are defined in MIL D 28000. Raster files are defined in MIL R 28002. The primary difference between the two is that raster files are like Polaroid images. They are essentially 'snapshots' of drawings or other documents. They can be edited by certain design tools, so they work well for the capture of legacy data. We raster scan old drawings, get raster files, add metadata (information in BI) and check the results into the PDM.

IGES files are a more 'intelligent' form of electronic data (called 'vector' data). IGES has the capability of transmitting two- or three-dimensional data (x, y, z), whereas raster is strictly two-dimensional.

I recommend that you scan military specifications for both types of files to get a better idea of the differences between them. It shouldn't take too long – about 110 pages for both documents, including charts.

We are using the raster version because it is my belief that raster is more universally accepted at this time, there are many raster viewers on the market, and we have had trouble transmitting some symbols in IGES format. We will probably transition to Adobe's Portable Data Format (PDF) in the near future for text documents.

Note that we are not only using raster files for legacy data capture. We are capturing native CAD files in our PDM for future changes and design re-use. When we check in the native files, the PDM saves the native version in one vault and the CM librarian converts a copy of an hpgl neutral file (generated by the CAD design tool) to raster format (using a utility in the PDM system) for view and distribution purposes and saves these raster files as separate objects (DI). Both file formats are related to the Design Document BI.

Question 8

Configuration audits. Are they really needed in this world of ISO/QA certification?

Answer 8

I believe that we still need configuration audits. FCA is needed to prove that we have met our design requirements, i.e., the design 'works'. PCA is needed to prove that we can build production units that meet the engineering drawings and that are physically and functionally the same as the prototype. Verification audits are needed to prove that we have cut in Major (Class 1) changes where we should have and to provide a spot check on the sourcing and manufacturing processes.

With the advent of PDMs, we need to check compliance to CM/DM/PDM procedures. It's not all automatic, yet. There will still be some procedure-driven CM activities or interfaces with the new PDMs for a while, at least. CM self audits aren't a bad idea, either. Nobody's perfect.

I think a significant portion of the time of CM personnel in the future will be spent planning, training and performing audits to assure compliance to the CMPs and proper operation of the PDM processes plus resolving CM-related problems. Fine-tuning of these processes will be another beneficial result of these audits.

Question 9

Why have a transition to production period when a prototype is already built? Why not have this period during the manufacturing of the prototype?

Answer 9

We need a transition to production period to 'pass the baton' from one set of performers to another. During development, engineering, CM, producibility, laboratory, and drafting personnel carry the ball. During production, manufacturing, quality assurance, sourcing and field support are the primary players. Many things have to happen during this period, as stated in the book:

- Verification of the TDP;
- Loading of the MRP system;
- Material ordering;
- Vendor interfacing;
- Project and process familiarization for production personnel;
- First article inspection;
- Change activity (formal internal control).

We are trying to get away from the old engineering 'hand-off' to manufacturing and encourage a smoother transition by 'teaming up' with our production brethren, i.e., no more 'pass the buck' or 'I did my thing – now it's your turn'.

Question 10

How will the Perry Initiative change the CM world?

Answer 10

The main thrust of the Perry Initiative is to utilize commercial CM practices more and rely less on military standards. I believe this should drive industry to develop 'best CM practices' for the development, production and support of our products. We should then negotiate these 'best CM practices' with our DoD customers to arrive at the optimum mix of cost versus control in each area of CM for each of our programmes and document the results of these negotiations in our CMPs.

We must be mindful that the Perry Initiative did not do away with all military standards. MIL STD 2549 will be a mandatory requirement upon most of us. It defines the data elements and electronic data delivery format/interface requirements for those of us who do business with the DoD. We will deliver our contract data (using PDMs) per this interface standard, and our government customers will retrieve and process these data using their CMIS, JCALS, JEDMICS (or other systems) on their side of the interface or they may just use the internet and utilize home pages on the WWW to access our PDM systems. In either case, however, we will have to meet the requirements of MIL STD 2549.

Question 11

Can you tell me what cost savings I will realize from a PDM system?

Answer 11

Your largest initial savings will be in reduced design data delivery costs. You will go from delivery of aperture cards and paper to delivery of CALS format electronic data (magnetic media or over the wire). Later, you will deliver 'in place' via your CITIS functionality.

You will also save via reduced CM activity cycle times and workforce reduction (you won't need a documentation reproduction and delivery organization, anymore). It is difficult to estimate cycle time savings. You can count heads in your documentation data centre, though. You may want to train one or two individuals who previously performed paper distribution related tasks to raster scan legacy data for conversion to files. You will realize significant savings in improved efficiency in all departments. For an exercise, just figure out what the savings would be for each 1 per cent increase in efficiency (productivity). Your PDM system will provide a significantly higher increase in efficiency than 1 per cent.

Question 12

How do I sell the idea of installing a PDM system to my management?

Answer 12

This is the toughest question I have been asked throughout my PDM project and as a result of writing this book. As in the case of all my answers, I will try to be truthful; however, in this case, I am concerned that my answer may be inadequate to your needs. Let's give it a try, though.

Management (mine, yours, everyone's) wants to hear how much money they will save by installing any new system. That is only right and reasonable. It is the correct business question, i.e., how much will I save, in what areas and when will I begin to realize these savings?

This valid question is difficult to answer at this point in the transition from paper to electronic data management because, simply put, there just aren't any significant amounts of historical PDM metrics data to analyze. It won't take long for your business to compile metrics on cost reductions in the data delivery area, i.e., from delivery of paper and aperture cards to electronic data delivery, but savings accrued from cycle time reductions and efficiency (productivity) improvements will take longer to measure.

The most important reasons for implementing a PDM system in your business are *control* and *compliance*.

Control

How many of your organizations have in place an effective system for the capture and control of your native CAD data files? Not many, I'd wager! The time you spend in trying to locate the latest version of your CAD files will approach zero when your PDM is in place and operating. Also, if you are like many of us who have design and drafting organizations that can't really tell 'which Revision B' file is the one I'm looking for (because Bob and Ray were both working on the same design and Ray is out on holiday today), you will rejoice when you find there is only one Revision B file in your PDM instead of multiple Revision B files scattered throughout your design network directories.

Compliance

The DoD has informed its suppliers that the only acceptable method for delivery of design data in the future will be via compliance to MIL STD 2549, Configuration Management Data Interface. This document received full approval on 30 June 1997, and will be included in RFPs as a required MIL STD for delivery of data on US government products. I think that it would be truly amazing to see a DoD supplier able to comply with the requirements contained in MIL STD 2549 without the functionality of a PDM system in place in their business. So, if you wish to continue to do business with the DoD, you would be well advised to implement a PDM.

These two reasons for implementing PDM systems, i.e., control and compliance, will, no doubt, go against the grain of many short-sighted managers (can you imagine 'Dilbert' trying to sell this idea to his manager?) but they are real, and those companies who choose to ignore them are incurring significant, and possibly fatal, risks.

Question 13

What would you do different if you were to start over (installing your PDM)?

Answer 13

I would try to do a better job of establishing realistic expectations earlier in the project in the following areas:

- PDM functionality;
- Integration of the PDM system into our programmes;
- PDM project cost;
- PDM system functionality and programme integration schedule milestones.

I had held a series of CM/PDM 'overview' meetings (two hours each) at the beginning of the project. I conducted these sessions first for mechanical and electrical lead engineers, then mechanical and electrical designers and draftspersons. Next, I moved on to quality assurance and manufacturing planning personnel, field service and logistics personnel and customer representatives. In other words, I focused on the users and our customers.

I should have included middle management early on in the process, for these are the people who make the funding decisions and who worry about integration and personnel training schedules. I was forced into an information broadcasting 'catch-up' mode at a time when all of my energies should have been focused upon coordinating PDM functionality implementation, programme integration and training.

If the expectation of middle management had been realistically established earlier in the project, I would have had more time when I needed it to implement and to integrate our PDM system. As you will see when you study the 'lessons learned' in Appendix D, there arose plenty of unplanned activities that ate up valuable time later on in the project.

I would also have started training earlier and installed PDM clients earlier.

Question 14

Can you provide a template for a PDM System Requirements Specification?

Answer 14

Yes. The SRS template is only a reorganization of the templates already supplied in this book.

Figures C.1 and C.2 plus the PDM-related sections of Figures 10.1 to 10.10 provide you with a list of topics to cover in your SRS. You should also re-read Chapter 9, Automated CM, paying particular attention to the sections on PDM Requirements Analysis and PDM System Requirements Specification. Next, read Appendix C, PDM tool and vendor evaluation and selection process. The information contained in these sections of 'Practical CM' provide you with what you will have to consider and include in your SRS inasmuch as these topics relate to your business.

Once you have generated your lists and completed your review of these sections, you should be able to create a 'strawman' (intermediate) SRS for your own business requirements. You can then 'flesh out' the details of your SRS as they pertain to your business processes and programme.

Question 15

What types of CM 'things' (effectivity date, release date, etc.) should no longer be part of the document because the PDM keeps track of these items now?

Answer 15

The master of the 'document' (in our old world, a paper or mylar drawing, specification, contract/non-contract data, proposal, etc.) will reside in your PDM system in the form of electronic data. If yours is an object-oriented PDM (and that is what I recommend), then your metadata (information about your design data files or office data files) will be represented by one type of object and will reside in a database. Native CAD design models and 'flat' files plus neutral files (for viewing and distribution) will be represented by another type of object. These objects will be related (another object) to each other and this combination of objects will be stored in vaults (directories on a UNIX network, for example) and will be accessed by all who need to use or view them (based upon a security access scheme). Engineering, manufacturing, CM/DM, etc. processes will be controlled by workflows within your PDM system to the extent you define them.

This methodology can be applied to control the cut-in of changes at effectivity points (dates, serial numbers or events) plus any other information that used to exist on the drawing and other documentation. Since all the information you need to capture, control, re-use, view, distribute, and implement your designs and engineering/manufacturing processes is available within your PDM system, you have no need for an external piece of paper containing effectivity and release information.

Question 16

Can you elaborate on the workaround you are using for the capture and control of your native CAD design model data until you have 'tight integration' between your design tools and your PDM in place?

Answer 16

I am currently encapsulating our Pro\Engineer and Mentor design files (no problems with two-dimensional 'flat' files) in order to assure that I have captured all associations, linkages and pointers related to our design models and engineering libraries for future design modifications and/or re-use. I use the 'tar' and 'zip' encapsulation methods. I then check the encapsulated files into our PDM native data vaults. I keep track of the contents of my encapsulated files via computer listings.

My goal for the future is, of course, establishing 'tight integration' where the designer or office worker launches their design or office tool (e.g. Pro\Engineer, Mentor, MSOffice, Interleaf) through our PDM tool system, works on their design or office tools during the day, with the PDM capturing and controlling the data via workflow processes, vaults, groups, roles, rules, and all the other 'goodies' provided by an 'object-oriented' toolkit type PDM.

Until industry provides the interface modules for each of these design and office tools, I am stuck with this rather clumsy and labour intensive encapsulation mechanism. This also means that I have to generate hpgl and/or milr raster (neutral) files for viewing by our third-party tools and for electronic distribution in CALS formats. I plan to implement full CITIS functionality as soon as possible but I can't do that until we get a few things (firewalls, internet solutions, external clients, etc.) established.

I am trying to follow the industry to stay current on interface module availability. Much is happening out there, especially with the large companies driving the PDM vendors for solutions. I have already learned that it's best to stick with well-established suppliers. I was bitten pretty badly by having some PDM customization (screens and image services module) created by an outfit who has stopped doing business in the PDM customization and tool integration arena. There are a lot of folks out there trying to get in on this rapidly moving market. I suggest that you carefully evaluate any potential vendors before you jump in with them.

Appendix C
PDM tool and vendor evaluation and selection process

Introduction

Mission

Your mission is to pick the right PDM tool, the right third-party software (view, print, format conversion, etc.) and the right companies and/or consultants to help you establish, integrate and support your new PDM system. My mission is to provide you with the right information, checklists and templates to help you accomplish your goals and to keep you from travelling down the wrong paths.

Scope

The scope of Appendix C includes a discussion of the PDM elements you should consider during your planning, evaluation and selection processes. It also provides two templates (Figures C.1 and C.2) for you to consult and fill in so that you will 'cover all the bases' and procure the right PDM tool/vendor combination to fit both your program and budget requirements:

Planning

Requirements analysis

If you have followed my recommendations in the introduction to Chapter 9, you have just finished reading that chapter and 'popped over' to this appendix. That means you have completed a first pass at your Requirements Analysis. Having done so, you are now prepared to study and complete the first three columns of the PDM tool evaluation and selection template presented in Figure C.1, 'Required', 'Desired' and 'Phase'. If, however, you chose to ignore my advice, you will probably have to go back and re-read Chapter 9 because this appendix and Chapter 9 go hand in hand, and unless your memory is a lot better than mine, you will need to refer back and forth frequently. I am sorry for this inconvenience but I still think this is the best way to present this information and my recommendations. If I tried to do it all in Chapter 9, I would have to go off on too many 'tangents'.

Scan through the PDM tools template in Figure C.1 now. You will see that most, but not all, PDM tool elements have been covered in Chapter 9. Those not covered in Chapter 9 are discussed in this appendix or other appendices.

Try to fill out the first three columns, 'required', 'desired' and 'phase'. Don't worry if you don't know all the answers. As I have previously stated, you will have to communicate with several people in order to achieve your CM/DM/PDM goals. The correct answers will become obvious as you progress through the PDM tool evaluation process.

Market analysis

The last two columns in the PDM tools template, 'out-of-box' and 'customization', require a significant amount of specific information about each of the PDM tools from which you will select several candidates for evaluation. You should apply 'grades' for each PDM tool when you fill out these columns. I suggest the following grading system:

1 Fully compliant with additional features;
2 Fully compliant;
3 Compliant with third-party tool support;
4 Non-compliant.

You can, of course, modify this grading system as you see fit to match your specific needs.

You should start by obtaining a comprehensive listing of the currently available PDM tools, their suppliers and CM/DM/PDM consultants. This is not a difficult task. All you need is access to the internet.

A good listing of CM/DM/PDM tools, vendors, consultants, seminars, books, and related materials is provided in the CM Resource Guide by Steve Easterbrook, the widely respected CM consultant and lecturer. He updates these listings on a regular basis. The CM Resource Guide can be viewed on the internet at the home page of the Association for Configuration and Data Management (ACDM): http://www.acdm.org. Another good source for CM/DM/PDM tools, vendors and related support products can be found at the CM *Yellow Pages* at URL: http://www.cs.colorado.edu/users/andre/configuration_management.html.

Once you have retrieved your listing of potential PDM suppliers, I suggest that you contact others who have embarked upon the PDM path. If you are employed by a large corporation contact your sister divisions or other folks you do business with. Many good things can be said about asking for help, especially in this area of endeavour. I suggest that you also subscribe to the CM Working Group mailing list. You will be exposed to a wealth of information. You can subscribe to the CM Working Group mailing list by visiting the quality web site on the internet at URL: http://www.quality.org/cgi-bin/majordomo. Easy as that! You will also have the opportunity to offer advice of your own. Groups like this and the ACDM are doing good things to enhance communication between CM folks like you and me.

I can't provide you with specific recommendations for tools or vendors in this book but I do participate actively in the CM Working Group on the internet, and have provided specific information about the PDM tool that my company uses as its Enterprise Data Management tool. I used the term Enterprise Data Management here, as compared to Product Data Management, because we have extended the scope of our Product Data Manager (PDM) systems at several of our businesses to include not only product data, i.e., specifications, design TDPs, QA test and inspection data, manufacturing planning data, product as-built data, but also internal procedures, proposals, contract deliverable and non-deliverable data and just plain old engineering design folders and things like problem reports and office tool products. The only business data not included in this category are those data controlled by our Manufacturing Resource Planning (MRP) System, Consolidated Purchasing System (CPS) and financial systems. We are currently in the process of defining the interface between our PDM and our MRP and CPS systems. We may integrate our PDM with our financial systems in the future.

Since today's PDMs have the power to effectively capture and control all kinds of data, why not use it? Enter data once – use many times! (Where have I heard that, before?)

PDM tool/vendor evaluation

Vendor demonstration

Your next step is to select five or six (you pick the number – the more the better) vendors who appear to have a PDM tool that may 'fit the bill'. Call them on the telephone, ask them a few introductory questions based upon your PDM tool and vendor templates, and if the answers appear satisfactory, ask them to send a representative to talk to you. Tell them that you want to see a demonstration of their tool (not a video clip – that usually indicates they are trying to impress you with 'smoke and mirrors'). Remember that a PDM system is complex, and it isn't very difficult for an insincere salesperson to rig up a phoney, or less than honest, demonstration. You will never know all the intricacies of what the PDM tool in question can do until you personally exercise all of its options after it has been customized to fit into your CM and other business processes. You can, however, learn a lot by making sure that the salesperson demonstrates each functionality that you have selected from the 'Required' and 'Desired' categories in the PDM tools template provided in Figure C.1 and the 'SOW task #' column in the PDM vendors template provided in Figure C.2.

Record your reactions to the results of each tool demonstration on the PDM tools and PDM vendors templates. Note that the PDM vendors template columns have different headings: 'Sow #'; 'Licensing'; 'Support'; 'Cost'; 'Rating'. The meaning of these headings will be obvious by the time you have finished reading this appendix. I suggest that you select a rating system similar to the following:

- A – Excellent and within allowable costs;
- B – Fully compliant and within allowable costs;
- C – Fully compliant but exceeds allowable costs;
- D – Non compliant.

You will get to be quite good at asking the right questions after a few demonstrations. Have the salespersons and technical representatives (if you can get them to come along) demonstrate how their tool functions so that it meets your criteria 'off the shelf'. If it doesn't, you may be in for some costly customizations that either your company or an independent tool integrator must create. It is important to probe deeply into the many facets of these PDM tools at the beginning. What may seem like a Rolls-Royce, with its slick screens and classy functions may turn out to be severely limited in functional expandability at the point when you really need to add some functionality or expand upon the out-of-the-box functionality. That is why I recommended an 'object-oriented' PDM in Chapter 9. There is more work involved up front in order to customize the product to fit into your operating processes but the more robust software engine and greater expansion capabilities will provide the flexibility you may need in the future.

Hands-on evaluation

The next step in your PDM tool evaluation process is to get in some 'hands-on' time. Until you get the 'feel' of the screens, navigation functions, workflows, queries, data check-in/check-out processes, plus administrative functions like user, client, host and vault creation, you won't know how well each tool fits into your business processes and network structure. I will keep bringing up the warning that what you don't find out now may hurt you later. Don't let the tool salesperson or technical representatives get away with telling you that 'this is only a demo, and we can't do this or we can't do that' – *hogwash*! You absolutely *must know the answers to all of your questions at this time*. If they have to contact their company's technical people and get back to you later, that's OK. Just make them do it! Remember that whatever their PDM tool doesn't do now, you will have to pay for specialized customization to accomplish later on, whether your company creates it or you have a customization house do it for you. See below. You will also have to worry about keeping that customized software synchronized with your PDM tool software as time goes on, and newer versions of the tool are rolled out.

So, with all that in mind, take the time to get your hands on the tool, have them loan you an evaluation version (I would worry about any company that refuses to do this), and run it through its paces. You won't be able to verify all of its functionality. Your network configuration and NFS software may preclude that, and you may not be able to set up a client/server environment during your evaluation period. However, you should be able to check out most of the basic functionalities specified in your PDM tools template.

Take plentiful notes, and mark up your templates. Record successes and failures. Count keystrokes required to perform common tasks. A difference of one or two keystrokes can add up to a considerable cost saving over a year or ten or twenty. You will probably be stuck with your choice for a PDM tool for a long time – to say nothing of the fact that many of your sister businesses (if you work for a large corporation) may jump on the bandwagon and implement the PDM which you so wisely selected. You have a wonderful opportunity to become either a hero or a goat at this point in your career. Take the time to check out these PDM tools thoroughly. Be a hero, not a goat!

Third-party tools

After you have selected a PDM tool, you must select whatever third-party software you need to support it in order to provide the overall functionality required of your PDM system. Functionalities such as PDM view (look at screen images of neutral files that represent designs or office tool products), casual view (see Chapter 9), file format conversion (for CALS deliveries and native-to-neutral file conversions), NFS programs (for client/server communication), on-line CCB conferencing software, cataloguing software, print software, red-line markup software and others may not all be incorporated into the PDM tool you select. You should try to pick a PDM tool that incorporates as many of these functionalities as possible, though, because if you have to procure them separately, you will have to integrate them into your PDM tool, possibly by customizing one package or the other or adding on a special customization package. Then, of course, you will have to worry about supporting all of this.

In any case, just follow the guidelines I established for your PDM tool, and you should be able to select the right software with minimal fuss.

Tool customization and integration houses

You still have a considerable amount of work to be done as part of your PDM tool/vendor evaluation task even after you have selected what you feel to be the best PDM tool set for your business requirements. As mentioned above, you may need to integrate multiple software products into a single system in order to implement all the required functionalities of your PDM. If you are extremely fortunate, the PDM tool which you selected can either be integrated directly with the third-party tool(s) you require or can be integrated with these tools by using an interface module supplied by the PDM tool vendor, the third-party tool vendor(s) or another software supplier.

If, however, you cannot match your requirements with off-the-shelf products, you must find some other way of providing customized software to perform the necessary integration function. Considering the often required UNIX to DOS conversions and the need to FTP files back and forth between servers, clients and staging areas, it is not unusual to find oneself in this situation.

Now, you have another choice. If your business has a software engineering organization, you may elect to take on this task yourself. Remember though, that you will be breaking new ground and, most likely, be trying to execute software code to accomplish different types of things from those your software people have developed their expertise in. You also will have to support this new customization and keep it synchronized with your PDM tool and third-party tool(s) as they evolve into newer versions.

An alternative to this dilemma is to contract the services of a PDM tool customization and integration house. Several of these 'newbies' to the CM/DM/PDM scene have sprouted up to fill this specific void. A few have successfully implemented PDM functionality in large industrial applications and, in so doing, have gained significant 'know-how' by jumping through the hoops and overcoming the challenges that would be new to the uninitiated. A lot of time and money can be saved by soliciting the services of someone who has 'been there, done that'.

Caution is the word of the day in this arena, also. Make sure you follow the same process in enlisting the services of a customization and integration house that you have employed in selecting your PDM and third-party tool vendors. You can utilize your System Requirements Specification (SRS) as the basis for your Statement of Work (SOW) and subsequent purchase order or contract with your chosen customization and integration house. Use your SOW to help you fill in the first column in Figure C.2, SOW task #.

An acceptance test plan should be generated and agreed upon by yourself and your customization and integration house during your contractual negotiations. You both need to be sure that, when they think they are finished, you have what you expected to have in the first place.

You should also assure that you have a plan in place to support any customized software furnished by your customization and integration house, whether or not they remain in business over the years. One way to guarantee this is to obtain the source code and sufficient documentation to allow you or another software house to maintain your customization code if your original service provider goes under.

Finally, carefully check-out the credentials of potential service providers.

Consultants

One indication of the longevity of this new world of automated, electronic CM/DM is the emergence of another classification of professionals – the CM/DM/PDM consultant. I won't, and probably couldn't, elaborate on where these folks came from but it's a sure thing that there is a need for their services in the minefields encountered by businesses trying to transition from paper to electronic control of their CM (and other) processes.

How do you know whether or not you need to employ the services of a CM/DM/PDM consultant? Well, my best advice is to first read through this book, fill out the templates (by then you will have communicated with the key personnel within your business), see several PDMs demonstrated (get in your 'hands-on' time) and talk to other sites and businesses (both within and external to your business) that have planned, implemented and integrated a PDM system.

If you still feel 'shaky', then talk to a consulting business and let them tell you what they can do for you. Demand specific information. Use your templates as the basis for your discussion and questions. If they can identify areas where they have experience that you feel is critical, then hire them to support those specific tasks. Let their performance be your guide for future contract awards. It could be that this initial meeting might indicate to you that you have the capabilities within your own business, which, in conjunction with this book, can enable you to proceed on your own.

On the other hand, the old adage of 'a stitch in time saves nine' might well prove to be the case for your business. A few good suggestions from a qualified consultant might steer you clear of the PDM 'minefields'.

PDM tool and support procurement

Now that you have evaluated and selected your PDM and third-party tools and decided whether or not you require the services of a consultant, you are ready to pull your internal and external team together, get your contractual documents in order and get started down the path towards automated CM.

Statement of Work

The requirements defined in your SRS plus the line items provided in Figures C.1 and C.2 can serve as the basis for your Statement of Work (SOW). This SOW should become an integral part of your contractual documents with all external parties whom you have chosen to support the implementation of your PDM system. Your SOW should be specific enough to leave absolutely no doubt in anyone's mind what the tasks and responsibilities of both sides are in this endeavour. You should be able to definitively answer the question, 'has this task been completed?' at the end of the performance period.

Deliverables should be uniquely specified – hours, material, products, systems (hardware, firmware and software). If a level-of-effort support procurement is embedded in your contract or purchase order, you should establish some means of accounting for time spent. The process should be auditable. Your SOW is your tool to assure you have received services and products that live up to your expectations. Remember that, when you sign off on satisfactory completion of all elements of your SOW at the end of the performance period, you are on your own.

Acceptance Test Plan

As mentioned earlier, you will need to conduct tests to determine whether or not your PDM system can successfully meet the requirements specified in your PDM System Requirement Specification. Only upon successful completion of a series of tests which demonstrate satisfactory performance should you sign off on your contract or purchase order, signifying completion of contracted efforts and deliveries.

Therefore, as part of your contractual negotiations with your suppliers, you should generate and agree upon an Acceptance Test Plan (ATP) which documents exactly what tests and procedures are to be conducted in order to demonstrate satisfactory compliance to your SRS requirements. Your ATP doesn't have to be a lengthy document. It should be clear, crisp and specific, with roles and responsibilities clearly defined. A good ATP can eliminate a lot of frustration and hard feelings at the end of the performance period.

Phased procurements

The tasks defined in Chapter 9 are grouped into four phases for implementation of your PDM functionality. The procurement of your PDM tool has to occur at the beginning of Phase 1 but the procurement of third-party tools, customization and integration services and additional licences is best accomplished in phases. The philosophy of 'divide and conquer' works quite nicely for this project.

The categorization of PDM functionality implementation by phase as described in Chapter 9 has worked well for my company. Though I think it will work as well for any company, you are free to make variations on which task fits into any phase. You may decide on more phases or less phases. It's up to you. I do recommend, however, that you try to quantify your PDM implementation and integration tasks so that you will have the functionality in place in the optimum time frame to meet your programme needs.

By splitting up the functionality implementation, you also can perform these tasks in manageable chunks, i.e., don't bite off more than you can chew. A lot of 'stuff' is going to happen all at once when you get rolling. Executing your tasks in a phased approach will help you manage your internal activities and your external subcontractors and suppliers more efficiently. You can also take advantage of 'lessons learned' from one phase to the next.

See Appendix D for examples of lessons learned.

Licensing provisions

An important consideration as you plan for your PDM system is the type, quantity and cost of the software licences you will need for your PDM software, your third-party software, your database software, your network software and your customization software.

Types of licences

There are really only two types of licences you will have to worry about for most of your software: 'concurrent use' and 'seats'. The number of concurrent licences you need depends upon the total number of PDM users accessing the system at any given time. This works out well if you have many potential users but they don't all sign on the system at the same time. It also allows you to install the software in as many PCs or workstations as you want. Just don't exceed the (contractually agreed upon) total number of users at any given point in time.

The ratio of total users to concurrent users will vary from company to company. My best information says a ratio of four or five to one is a reasonable number. I would recommend that you procure less licences than you think you will need (try 80 per cent) and then buy more when you hit your limit. Software suppliers are always ready to take additional orders.

This advice applies to all software that comes with the concurrent licensing option, and in my experience, most does.

Single user licences or 'seats' are just that. One licence for each PC or workstation. This also means that the maximum number of PCs and/or workstations you can install is limited by the number of seats you have purchased.

The decision is yours. It's essentially a trade-off between cost and flexibility. Calculate your costs for different scenarios. Figure out what is best for your business.

Cost

If you are employed by a large corporation, then you either already have in place corporate agreements with your potential PDM and third-party software providers or you are in an excellent position to negotiate such an agreement.

In either case, try to strike a bargain with your software supplier(s). They are just as hungry as the companies that employ us. You have everything to gain and nothing to lose. The key words here are negotiate, negotiate and negotiate. And don't forget to include the software maintenance costs in your negotiations. Also, ask for special 'deals' such as free training and integration support, if only for a limited time. Take everything you can get! Whatever you don't get during your negotiations, you will pay dearly for later on.

Long-term support

You will need to consider the following support elements in your planning for maintenance and possible future modifications, including functionality enhancements to your PDM system:

1 Internal support;
2 PDM tool support;
3 Third-party tool support;
4 Customization supplier support.

Make sure you discuss these elements of support with your own personnel and your potential suppliers. Don't forget that each vendor's products must 'play' with your other PDM or third-party tools as each evolves into newer revisions. Try to place the burden of compatibility guarantees upon your suppliers.

Make sure that you get a well-integrated support package negotiated into your purchasing agreements. Remember, it's going to be a case of 'pay me now or pay me later'. Get your best deals up front while you are still in a position to bargain.

Also, it is absolutely mandatory that you and your suppliers reach an agreement at the time of negotiations on specific roles and responsibilities for yourself, your employees and your supplier's employees. You don't want to find yourself in the position of having to argue with your supplier(s) about 'who does what' during the course of their period of performance. You can avoid this dilemma by clearly defining roles and responsibilities up front and by documenting your agreements in your SOW.

Finally, make sure you get your customization supplier to agree to provide you with the source code plus adequate documentation for any software they supply. You will need this package to support your customization software if your supplier ever goes out of business or if you decide that you can do the job yourself.

Good luck with your PDM tool and vendor evaluation and selection.

Topic: PDM tools

PDM tool element	Required	Desired	Phase	Out-of-box	Customization
Basic functionality:					
Check-in/checkout					
View:					
Images					
Markups					
Composites					
Vu-lists					
Vu-sets					
Print					
Raster edit					
Distribution					
File format conversion					
Legacy data conversion					
On-line CCB:					
Red-line markup					
From–to composites					
Mail notification					
Routing lists					
Workflows					
Voting options					
Queries					
Reports					
Import/export					
User access security					
Casual view					
List/icon/tree views					

Figure C.1 – PDM tool evaluation and selection template

Topic: PDM tools

PDM tool element	Required	Desired	Phase	Out-of-box	Customization
Design tool interfaces:					
Pro\Engineer					
Mentor					
SDRC					
Flat file tools					
System engineering tools					
Other design tools					
Office tool interfaces:					
MSOffice					
MSProject					
Interleaf					
Other office tools					
MRP tool interface					
CPS tool interface					
Operator interface:					
GUI					
Drag and drop					
Screen navigation					
Screen dynamic refresh					

Figure C.1 Continued

Topic: PDM tools

PDM tool element	Required	Desired	Phase	Out-of-box	Customization
Objects available:					
Business Items					
Data Items					
Relationships					
Vaults					
Groups					
Processes					
Rules					
Roles					
User					
Host					
Resource requirements:					
Workstation					
PC					
MacIntosh					
Network					
NFS package					
Documentation:					
On-line					
Manuals					
Licensing agreements:					
Concurrent user					
Single user seats					
Site					

Figure C.1 Continued

Topic: PDM tools

PDM tool element	Required	Desired	Phase	Out-of-box	Customization
Maintenance:					
Screen create/modify					
Administration					
Customization					
Training:					
Vendor provided					
In-house					
Overview					
Hands-on					
Third-party tools:					
View/markup					
File format conversion					
Workflow					
Print					
Backfile conversion					
Distribution					
Change process					

Figure C.1 Continued

Topic: PDM vendors

Vendor work element	SOW task #	Licensing	Support	Cost	Rating
Basic functionality:					
Check-in/checkout					
View:					
Images					
Markups					
Composites					
Vu-lists					
Vu-sets					
Print					
Raster edit					
Distribution					
File format conversion					
Legacy data conversion					
On-line CCB:					
Red-line markup					
From–to composites					
Mail notification					
Routing lists					
Workflows					
Voting options					
Queries					
Reports					
Import/export					
User access security					
Casual view					
List/icon/tree views					

Figure C.2 – PDM vendor evaluation and selection template

Topic: PDM vendors

Vendor work element	SOW task #	Licensing	Support	Cost	Rating
Design tool interfaces:					
Pro\Engineer					
Mentor					
SDRC					
Flat file tools					
System engineering tools					
Other design tools					
Office tool interfaces:					
MSOffice					
MSProject					
Interleaf					
Other office tools					
MRP tool interface					
CPS tool interface					
Operator interface:					
GUI					
Drag and drop					
Screen navigation					
Screen dynamic refresh					
Objects to be provided:					
Business Items					
Data Items					
Relationships					
Vaults					
Groups					
Processes					

Figure C.2 Continued

Topic: PDM vendors

Vendor work element	SOW task #	Licensing	Support	Cost	Rating
Rules					
Roles					
User					
Host					
Resource integration:					
Workstation					
PC					
MacIntosh					
Network					
NFS package					
Documentation:					
On-line					
Manuals					
Licensing agreements:					
Concurrent user					
Single user seats					
Site					
Maintenance:					
Screen create/modify					
Administration					
Customization					
Third-party tools:					
View					
Format conversion					
Workflow					

Figure C.2 Continued

Appendix D _____
PDM planning and implementation lessons learned

Introduction

My goal for this appendix is to provide you with 'lessons learned' by my company and myself over the course of the time spent during the process of planning, implementing and integrating our PDM system. I want to be as accurate as possible so that you are presented with meaningful information but I don't want to divulge any company proprietary information.

I will be specific except in those cases where to do so would cause me to divulge information of a proprietary nature. There aren't too many of these cases, and I don't think anything of significant value will be lost by their omission or by the generalizations which I have substituted for the specifics in these particular instances.

The following list of lessons learned topics is provided to direct the reader to specific topics of interest.

Lessons learned topics

Lesson 1   Distributed versus non-distributed networks
Lesson 2   PDM user resources
Lesson 3   PDM user training
                   Overview
                   Focused
Lesson 4   Design/office tools – tight integration with PDM
Lesson 5   Legacy data
Lesson 6   UNIX versus DOS issues
Lesson 7   PDM system operating speed
Lesson 8   Image services
                   View lists and view sets
                   Markup overlay files
                   'From/to/delta' composite files

Lesson 1 Distributed versus non-distributed networks

Be careful to indicate that you want a distributed database when you install your PDM software, whether or not you currently plan to use a single or distributed database in the future.

Risk if lesson not heeded

You may have a difficult time transferring metadata from your non-distributed database to a distributed database if, in fact, the time does arrive when you need a distributed database.

Discussion

When we installed our PDM tool software, we installed it as a non-distributed database. This established database table space allocations for a single server.

We later transitioned to a newer version of our PDM tool software, and learned that if we had indicated 'distributed' versus 'non-distributed' during the initial installation, we would have established multiple table spaces versus the single table space that we established for our non-distributed database. We had, therefore, locked ourselves into a position where it was easier to re-enter our metadata than to try to reallocate it from our single table space to the newly created multiple table spaces. Files (Data Items) don't present a problem. Metadata (Business Items) do.

Lesson 2 PDM user resources

Plan and obtain funding for your user and client resources early in your PDM implementation programme.

Risk if lesson not heeded

Your PDM functionality may be in place and ready to use before your client resources (PC/workstations) are in place and ready to use.

Discussion

We have been increasing our PDM functionality in a phased approach and, in parallel, we have been integrating the PDM into our existing programmes. Even after having conducted several PDM training and overview sessions during which the PDM 'story' was broadcast, it has been difficult to establish hardware resources (PCs and workstations) as PDM clients in as timely a manner as needed to establish users for on-line CCBs and design baseline capture and control. Network issues and overhead funding for resources have a significant impact on the process of establishing PDM users and clients.

Be sure to plan this activity well in advance of achieving PDM functionality. Make sure that you have 'buy-in' from those responsible for providing the necessary funds to accomplish your PDM goals.

Lesson 3 PDM user training

Start PDM user training early in your PDM implementation programme.

Risk if lesson not heeded

PDM functionality and client resources (hardware and software) may be in place and ready to use before your users are trained.

Discussion

Your PDM user training plan and the necessary facilities and materials to support it will take some time to develop and put in place. If you do not start this activity early in your PDM implementation programme, you may find it necessary to provide individualized training for your CCB members, designers, draftspersons, CM personnel and others. In our case, it was necessary to provide this training on employees' personal time since our overhead funding for PDM training was cut prior to the start of the training programme.

You should start training users as soon as possible. My experience indicates that six to eight hours per student is a good rule of thumb, at least to get them started off in the right direction. You can provide additional, specialized training for third-party tools (e.g. view, markup, file format conversion, CALS distribution, backfile conversion of legacy data) later, if necessary. Training should be conducted in a training facility with approximately ten PCs or workstations.

Two categories of training are recommended, as indicated in Chapter 9, overview and focused.

Overview training

The overview training should cover basic PDM theory, e.g., objects, relationships, navigation, screens, menus, view, markup, distribution and queries. Instructor demonstration should serve to convey information presented in training materials. Training manuals, with table-oriented, step-by-step procedures should be supplied, discussed and worked through by both the instructor and the trainees.

Focused training

The focused training should be directed towards the specific task(s) to be performed by groups of employees, e.g., CCB members, designers, drafts-persons, customers and management personnel. This training should cover such things as workflows, check-in/check-out procedures, view, markup, etc., and include hands-on sessions involving actual production processes, question and answer sessions, homework tasks plus discussions on how to obtain help 'on-line' and from training materials, user manuals and your company's 'help desk'.

Make sure that your students' client PCs and workstations are ready for them to use when they complete training, or they will forget what they just learned soon after they walk away from your training room. Your PDM project will definitely lose momentum if these resources are not in place when required.

Lesson 4 Design/office tools – tight integration with PDM

Know your PDM capabilities and limitations with regard to capturing mechanical and electrical design model data. Complex mechanical and electrical designs (models) must receive special attention for induction into PDM vaults.

Risk if lesson not heeded

You may not be able to re-use or modify your design model data in the future.

Discussion

The process of capturing flat or two-dimensional files from non-model producing design tools is straightforward. It is relatively easy to check these files into a PDM, with or without going through staging areas. The process of capturing design files from the more complex, model-producing design tools is, however, a different issue.

We developed a workaround for Mentor and PRO\Engineer files (encapsulation) which we plan to use until the industry produces a PDM-to-design-tool interface module for each design model tool. These interface modules must provide 'tight integration' and enable the CM/DM control plus the engineering process control required during development. We are now, however, limited to capturing baselines for individual designs following development (at the point in time when we usually capture the development baseline on paper drawings). Additional detail on this topic is presented in Appendix B, Question 16 and in Chapter 9.

Be careful to identify the tool set used by your organization to generate both your legacy data and your current and future CAD data early on during your transition to your PDM system. This ensures that your PDM system functionality is available and roles and responsibilities are determined and assigned and procedures are in place when you want to start capturing your electrical and mechanical model design data.

Lesson 5 Legacy data

Use caution when selecting a methodology for the capture of legacy data.

Risk if lesson not heeded

The 'value-added' versus the 'cost-to-implement' equation can easily go in the wrong direction if a thorough analysis of your current and future programme needs is not conducted prior to taking action.

Discussion

We all have a choice when it comes to addressing the subject of how we are going to capture our legacy data or if we are going to capture it at all and check it into our PDM system. We have the option of leaving it as is, i.e., paper and aperture cards plus various files from many different design tools, some current and some obsolete.

We chose to raster scan aperture cards for our mature programmes. So far, this has worked out well. The process is relatively inexpensive, and using backfile conversion utilities, the appropriate metadata has been automatically entered into our PDM database.

This method appears to be more efficient and less expensive than trying to capture the old CAD files and maintain the hardware and software resources used to generate them. Normal drawing maintenance can be achieved through the use of a raster editor as future changes are approved.

Capturing legacy data for CALS deliveries

The process of capturing legacy data images in raster file format, loading metadata which describes these files, relating the resulting objects which represent these files within your PDM, and checking these objects into your PDM vaults is an incremental and iterative process. We procured 'backfile conversion' utilities to automate this process.

Utility 1 plus a UNIX shell script first renames the scanned MIL-R-28002, Type 1 raster image files using the following standard naming convention:

1234567_001_A.ras

Where:

'1234567' is the base drawing #
'001' is the sheet #*
'A' is the revision
'ras' is the assigned extension (raster)

*One file per sheet is the convention we used. One exception to this strategy is A-size drawings, which often have four images per aperture card (sometimes referred to as '4-ups'). Since we obtained our raster images by scanning aperture cards, we inherited an occasional four sheets (images) per file versus one sheet (image) per file. *Note*: This condition (4-ups) is unnacceptable for some programs.

Utility 1 creates Design Document Business Items (BI) from metadata contained in index files which were generated during the scanning process. These index files contain the holorith data punched into the scanned aperture cards. These data are then converted into 'Design Document' BIs. See Figure 9.2 and Figure 9.3.

Utility 2 registers the renamed files with the PDM database, thus creating Data Items (DI). This process makes these MIL-R-28002 raster files known to the PDM. This same utility then relates the DIs to the BIs and checks the combination into the PDM vault.

Utility 3 transfers ownership of the DIs to the vault, assigns them to a project, and finally adds a title to the Design Document BI.

When all files for a program are inducted into the PDM, errors and problems resolved and revisions validated against the current 'Record of Authority' database, the files and metadata are 'baselined'. This process further restricts access to the files and initiates specific PDM functionality. This functionality varies among PDM systems but generally involves the mechanism for checking in and out the files for update (DI) and the revision change process (BI). Once the legacy data are baselined, you can initiate your on-line CCB activities.

Note: You can always enter your metadata via keypunch but the above process is much more efficient for capturing and baselining legacy data.

Lesson 6 UNIX versus DOS issues

UNIX/DOS interface issues add complexity and cost to PDM implementation.

Risk if lesson not heeded

PC clients trying to access files stored in your PDM UNIX vaults may not interface properly with your network UNIX environment resulting in inability of your PDM or third-party tools to transfer files to staging areas for viewing, printing, distribution, conversion, or markup.

Discussion

Your PC PDM clients and/or your PC 'casual view' tool may require additional customization and/or third-party software to function properly. Also, security issues (UNIX permissions) must be addressed when files are copied or checked out of vaults owned by your PDM tool to UNIX and/or DOS 'staging areas' or work locations for view, print, markup, distribution, and/or conversion in order to provide access to these functions for the appropriate personnel. In other words, you don't want to allow everyone access to all directories in your PDM system but your PDM and casual view tools must be able to access the files. The flexible, highly controllable security access functionality of your object-oriented PDM is one of its most important functions.

Plan your permission and network structure carefully so that the right people and only the right people have access to your data, and so that your PDM and casual view tools can do their jobs.

Another DOS issue is the filename length limit. UNIX can have long filenames. Your design tool file naming conventions will demand long filenames. DOS is limited to a root filename of eight characters and an extension of three characters. You can overcome this problem by installing Windows 95/98 or Windows NT on your PCs.

Lesson 7 PDM system operating speed

Use appropriate NFS software for your PDM PC clients.

Risk if lesson not heeded

PC clients may be unacceptably slow if connected to your server(s) via an inappropriate NFS software package.

Discussion

We started implementing our PDM system using the wrong NFS package installed in our PC clients. The amount of time required for typical PDM actions (e.g. sign-on, query, check-in, check-out, workflow operations, view, markup and print) was unacceptable slow. We then evaluated various NFS packages and finally chose one which had no detrimental effects on any of our installed PC software base. This new NFS software package allowed for variable timeframes for transfer of data packets as compared with the fixed timeframes allotted for transfer of data packets by our original NFS software. The result was orders of magnitude shorter wait times for PDM system operation.

I have heard that Windows 95/98 and Windows NT also speed up operating times for our particular PDM tool. I don't have any information about their affect on other PDM tools but I am mentioning this so you can at least ask the question of your potential PDM tool suppliers.

This is also another area where it doesn't hurt to ask around for advice. Start within your own company and then ask on the internet. Also press your PDM tool vendor for information. This area is one of the 'hot spots' in the world of PDMs. Nobody wants a slow system.

Lesson 8 Image services

Learn the specific details of how your view and annotate (markup) tools work.

Risk if lesson not heeded

You may find yourself in the position (as I did) of having entered all of your legacy data into your PDM vaults and not being able to view images and markups, plus route design change packages around to CCB members for review and approval *in the expected manner*. You may also not be able to automatically attach markups that you have created to the various objects (change notices, design documents) to which they should be attached. The specific PDM functionality to focus your attention on up front are the mechanisms for handling:

- View lists and view sets;
- Markup overlay file 'creation' and 'save' mechanisms;
- 'From–to' image comparison file 'creation', 'save' and 'routing' mechanisms.

Discussion

We are using the view/annotate (markup) tool of a major supplier of such tools. We also employed the services of a PDM tool customization vendor.

View lists and view sets

I had generated a System Requirement Specification (SRS) for our PDM system which defined the specific requirements for each area of functionality, e.g., view, markup and CCB process, but this SRS did not get into the minute details of the design which would provide the functionality necessary to realize these capabilities.

As it turns out, neither did our customization vendor, until late in the programme. The problem was, as I discovered after we had checked all of our legacy data into our PDM vaults, that there were such things as 'view lists' and 'view sets' embedded in the third-party view/annotate software that we had chosen, purchased and installed in our PDM system.

In order to be able to view drawings in their entirety (i.e., call up a drawing, including all its sheets) and view them sequentially (as compared to one file (one drawing sheet) at a time), I learned late in the process that I would have to create view lists for each drawing. A view list is just what it sounds like – a list of drawing sheets (files/images) to be called up and viewed by the view tool. Also, in order to automatically capture and attach markups and from–to composites to the appropriate objects, I would have to create view sets. A view set is like a folder that captures and collects all files with the same root filename (same characters to the left of the decimal point).

If I had known about view lists and view sets before we inducted our legacy data into our PDM system, I would have been able to insist that our PDM customization vendor incorporate the appropriate functionality into our backfile conversion program, i.e., the program that took our legacy data files, renamed them, created Business Items and Data Items for them, related them to each other, and checked them into our PDM vaults.

Of course, the customization vendor should have been the one to determine what was needed to accomplish the functionality that they had signed up to provide but that fact did not help after it was too late to do anything about it. The responsibility was mine to get the job done right. This is another reason why you should carefully follow the recommendations provided in Appendix C regarding your dealings with PDM customization vendors.

The end of the story goes like this: for our legacy data, we have to manually create view lists and view sets (a series of keystrokes) when we use these data to support a change package submission to our CCBs for approval. This process could have been automatic instead of manual.

Markup overlay files

The problem with the creation and saving of markup overlay files is that I learned late in the game that the process we were using for initiating the network dispatcher and multiplexer were different from the PDM tool manufacturer's recommended procedures. This variation in startup techniques limited markup file creation and 'save' functionality. Once we learned how to properly initiate the dispatcher/multiplexer sequence, everything worked fine. We had wasted a lot of time troubleshooting this problem before we discovered (with the help of our customization vendor) what was wrong.

'From/to/delta' composite files

I had wanted the change package that would be submitted to the on-line CCB process to contain a composite file that would illustrate the proposed change as a coloured version of a 'to' image (the proposed design) overlaid on a black and white 'from' image (the current approved design), with the difference shown in a third colour. The salesman for the third-party image view tool assured me that their product contained this functionality. Well, as it turned out, after the software was purchased and installed, it did not work as promised. I could select the 'compare' option on the view tool and manually set up the files for comparison but I could not then 'save' this composite file and route it around to each CCB member for review. Each CCB member would have to set up the composite file for themselves.

I complained to the tool salesman. He complained to his technical director. The technical director thought it was a good idea and would put it on the list for a future functionality update for the product. Great! In the meantime, we have a few more keystrokes to deal with.

So, watch out! As I have said over and over in this book, make sure that any salesperson you deal with *proves* to you that their product is all that it is represented to be.

As a final note, you should think about how you want to select images to be viewed. We have four options: single drawing menu pick, multiple drawing menu pick, drag and drop (single images), and double click (single images). You must decide which method or combination of methods works best for your business processes.

Lesson 9 Production and training environments

Create a training environment for your PDM training activity.

Risk if lesson not heeded

You or your PDM students may corrupt your production data.

Discussion

You will have worked hard to capture your legacy and current design files, add metadata to them and check them into your 'production' environment. You should play it safe by creating a separate 'training' environment to use for training your future PDM system users so that you don't risk corrupting or losing your production data.

Lesson 10 Casual view

Make sure you understand the level of difficulty and complexity involved in implementing this functionality and integrating it into your business processes. Document all aspects of the task in your Statement of Work before you execute a contract with your supplier.

Risk if lesson not heeded

You may find yourself technically unable to use your casual view functionality on one or the other (or both) of your UNIX or DOS environments, and you may be limited in the extent to which your security access regulations permit you to use this viewing capability on your programs.

Discussion

You have three issues to deal with in implementing your casual view functionality:

1 Search engine;
2 View tool;
3 Security access.

The purpose of your casual view functionality is to allow many users to view and print (if necessary) drawings and documents you have captured and inducted into the controlled vaults of your PDM system with a minimum of expense (inexpensive view tool versus costly PDM user licence) and fuss (less key strokes required to view an image). The view tool is 'cheap' and easy to find but the search engine may be more of a problem to locate, build and integrate. We bought a 'package deal' from a customization house which sounded good at the outset, However, we discovered limitation upon limitation as time went by. Also we learned (in time) that, in order for each user to be exposed to the same security access rules as our PDM users, each must have an account on our UNIX network servers. We opted for 'open' user access, i.e., no security access for specific program documents rather than deal with the administrative burden required to limit access. We should have explored all of our utilization scenarios and discussed them with our supplier before we signed up for this 'package deal'.

Now, don't take me wrong. The casual view tool is a great idea, and it's the way to go for this functionality in a large, project-oriented corporation. You must simply assure that all the cards are laid out on the table up front and that everyone understands the work to be done and the non-recurring cost to be borne. Casual view functionality is an appropriate supplement to your PDM view functionality and may fill the bill for your organization but don't get taken in by a slick salesperson. Make them demonstrate, in your environment, what this casual view package can do for you. Also, even if they claim to have implemented casual view at their facility (and are willing to give you a demonstration) be wary. A single installation does not a product make. Also, include a requirement for user and maintenance documentation for this product in your SOW. Finally, determine up front how you (or the salesperson) is going to support this product in the future.

Lesson 11 Customer approval signatures (raster images)

Your customer may require that you provide as part of your PDM functionality certain capabilities which they feel are needed to do their job and not be exposed to unacceptable risks. The customer may offer to foot the bill for these special requests but you must be ready to accept the impact to your PDM schedule in order to accommodate the integration of these additional capabilities.

Risk if lesson not heeded

You may suffer unacceptable impacts to your schedule.

Discussion

In our case, this request came in the form of a requirement to have our PDM system affix a raster image of the customer representative's signature or initials upon the document in question immediately after they indicated approval (via keystroke) of a new design (signature) or a change (initials) to a previously approved design, with no human intervention occurring, either on their part or on our part.

To implement this functionality, we conducted a market survey for software which would perform the stamping function and which would be compatible with our PDM tool. It wasn't difficult to find that software. What was difficult, however, was jumping through all the hoops necessary to get this functionality to work.

The task elements to be dealt with in order to provide this raster signature functionality were:

1 Obtain customer approval personnel's signature and initials (paper).
2 Scan these signatures and initials to create raster image files.
3 Re-size (experimentally), invert (mirror) and convert these files to uncompressed hexadecimal files.
4 Generate a stamping script using the basic commands of the newly procured software package.

5 Generate PDM tool customization to integrate this stamping script into our PDM tool workflows for document approval and on-line CCB approval processes. *Note*: This activity required the services of a customization house.
6 Generate a matrix showing the pixel count to the proper location for the stamp images on all drawing sizes for both signatures and initials.
7 Proof out the functionality in both test and production environments.

Each element described above was not in itself a difficult task but the combined steps were quite time-consuming, especially since many of the tasks were of a 'hit or miss' nature.

This was one more lesson in the wisdom of understanding the true scope and depth of the task before agreeing to incorporate additional functionality in the critical path of the PDM integration schedule.

Lesson 12 Creating new screens and workflows; modifying existing screens

Make sure your personnel develop PDM tool maintenance skills early on in the project.

Risk if lesson not heeded

You may have to contract out for the services of a PDM tool customization and integration vendor for simple tasks that your own people should be able to perform.

Discussion

We sent personnel to training classes offered by our PDM tool supplier. Our customization vendor did, also. Both groups of individuals were exposed to the same training programme. Our customization vendor's personnel returned to their plant and continued to study the material and experiment with sample cases until they were proficient in screen creation and modification and process troubleshooting. Unfortunately, our personnel went on to address the issues of the day and did not develop their PDM skills.

We have paid for our resulting lack of expertise many times over both in cost (to repeatedly retain the services of our PDM customization house) and in lost time (schedule extensions).

My point here is, obviously, to train your personnel by exposing them to the proper courses and then to follow up to assure that they study, experiment and use the material presented to them during these initial classes *plus* put in additional time when they return to their jobs to *really* understand how your PDM tool works.

Actually, to be fair to our information services organization, we now have a network/system administrator who demonstrates the level of ownership required to do what has to be done, not only to support and maintain our installed PDM system, but who is in the process of learning how to modify the PDM tool code (with the concurrence of the PDM tool vendor) plus the customization code. This is the kind of dedication that you should be looking for in your information services personnel. It takes a lot of time and a high level of tolerance to frustrating events to be successful in this difficult job. However, it is mandatory that you find this kind of individual to support your PDM project.

Your business will be utilizing your PDM tool for a long time. You must have personnel within your organization who can maintain it and modify its functionality as required.

Index _____